GEOLOGY *of* BRITISH COLUMBIA

GEOLOGY
OF BRITISH
COLUMBIA

A Journey through Time

NEW EDITION

Sydney Cannings

JoAnne Nelson

Richard Cannings

GREYSTONE BOOKS

Vancouver/Berkeley

Greystone Books Ltd.
www.greystonebooks.com

Cataloguing data available from Library and Archives Canada
ISBN 978-1-55365-815-3 (pbk.)
ISBN 978-1-55365-816-0 (epub)

Editing by Nancy Flight
Copy editing by Iva Cheung
Front cover photograph © Michael Melford/National Geographic/Getty Images
Back cover photograph by JoAnne Nelson
Cover and text design by Naomi MacDougall
Maps on pages 53, 58, 61, 84, 88, 116 and 118 by Eric Leinberger
Maps on pages 10, 16, 25, 26, 28, 29 and 43 by Maurice Colpron
Photo on pages ii–iii: A small band of Bighorn Sheep take refuge on the basalt cliffs of the Chasm, north of Clinton.
Printed and bound in China by 1010 Printing International Ltd.
Distributed in the U.S. by Publishers Group West

We gratefully acknowledge the financial support of the Canada Council for the Arts, the British Columbia Arts Council, the Province of British Columbia through the Book Publishing Tax Credit and the Government of Canada through the Canada Book Fund for our publishing activities.

Greystone Books is committed to reducing the consumption of old-growth forests in the books it publishes. This book is one step towards that goal.

Contents

Preface

MOST PEOPLE ARE drawn to the natural world, at least at first, by its spectacular beauty. We are drawn by the spiritual majesty of mountains, by the perfect elegance of turquoise lakes set in sculpted cirques, by rainbowed waterfalls plunging over rock cliffs, by smooth rock bluffs perched next to an ocean sunset and by rivers winding through dramatic canyons. Whether we realize it or not, we are drawn to rocks and erosion. But many of us are easily distracted by the flashing wings of birds, the colourful beauty of wildflowers or the nobility of large antlers and forget to inquire into the postcard scenery behind them. This scenery is the story of rocks and erosion. But although the stories of the formation of British Columbia, the rise and fall of its mountains and the sculpting of its valleys are fascinating on their own, they are not separate tales from those of the flashing wings, the colourful wildflowers and the big antlers. This book links these stories.

Geology is the foundation of natural history. It tells the story of the minerals that form the soil in which flowers, grass and trees grow. It tells of the movement of continents, the rise and fall of mountain ranges, and the creation of ocean and lake basins, all of which determine the climate of a region, which in turn determines what plants and animals can live there. Continental movements separate and then rejoin biotas, or the flora and fauna of regions, over many millions of years, and glacial ages do the same over tens of thousands of years. Mountain ranges and great valleys determine

how animals and plants can or cannot disperse over the landscape. It is impossible to understand ecology and evolution without entering the realms of geology and climatology.

This is an exciting time for geologists and naturalists in British Columbia. The geological history of the province is being rewritten constantly—many discoveries have been made and many new theories have emerged in the decade since this book was first published. And in the realm of biological geography, advances in genetic technologies have shone a bright light on the recent history of animals and plants. Drawing on these latest advances in knowledge, we have completely revised our earlier book to produce this edition. We are extremely fortunate to be joined in this latest effort by our friend JoAnne Nelson of the BC Geological Survey, one of the leaders in big-picture geology in northwestern North America.

In this book we attempt to set the stage for the stories of the natural world around us today. We ask questions such as: How were these rocks formed? Where did they come from? Have there always been mountains here? Where did the Ice Age glaciers come from and what did they do? Where did all the animals and plants go when Canada was covered in ice? How did they get back here? Why does British Columbia have such a diversity of climates and ecosystems? And what are the patterns in those ecosystems?

This is by no means the definitive book on the geology of British Columbia. Rather it is an attempt to bring together a brief, understandable history of the province's geological features and the history of its living creatures into one cohesive story.

Most outdoor enthusiasts want to learn more about the world they walk, hike, bike, canoe, kayak or climb through. Until recently there were no books in British Columbia that filled that need beyond the level of the generic field guide. This book is designed to both entertain and educate nature lovers of all kinds, and we hope that it will inspire them to want to know more.

The red volcanic rocks of the Rainbow Range in Tweedsmuir Provincial Park, the product of a hot spot in the earth's mantle, are a testament to the dynamic nature of British Columbia's geology.

Acknowledgements

THIS BOOK WAS inspired by many years of conversations along forest trails, around campfires and in university coffee rooms and government offices. The naturalist tradition in British Columbia is largely an oral tradition, and many knowledgeable people have freely shared their stories with us. We could not have begun to write this book without them.

Jim Monger, father of terrane theory in the interpretation of the Cordillera, has been the well of understanding and inspiration to us all. Maurice Colpron of the Yukon Geological Survey freely shared his knowledge and ideas of terrane evolution and created all the terrane maps, keeping them simple without sacrificing understanding and precision. Eric Leinberger expertly composed the other maps. Others that gave freely of their knowledge were Dennis Demarchi, Carlo Giovanella, Richard Hebda, Catherine Hickson, Rick Kool, Don McPhail, June Ryder, Howard Tipper and Doug van Dine. Their comments and suggestions greatly improved the depth and accuracy of the book. Any remaining errors, however, are ours alone. Trevor Goward, Margaret Holm, Leah Ramsay, Douglas Leighton, Nancy Baron and David Stirling all gave valuable suggestions on the content and style of the original chapters.

Donald Gunn crafted most of the line drawings; other illustrations were provided by Bob Carveth and David Budgen. A number of excellent photographers contributed their images—we would especially like to thank Chris Harris, Jared Hobbs, Ken Wright and

4

Mark Hobson in that regard. Rolf Ludvigsen and UBC Press kindly gave permission to use the photographs of fossils from their fine book *Life in Stone*; the photographs were provided originally by the authors of the chapters of that book: Andrew Neuman, Elisabeth McIver, James Basinger, Mark Wilson, Ruth Stockey and Wesley Wehr. Other images were provided by Bob Turner, Jeff Mottershead, Jim Baichtal, Jim Pojar, Richard Hebda, Robert Cannings and Scott Webster.

Introduction

As boys we poked into a lot of natural history—we looked at birds, we caught trout in small creeks, we helped our older brother make a butterfly collection, and—for a short time at least—we kept a leaf collection. But we poked a bit into geology, too. As small boys we looked wide-eyed at the perfect twigs of a dawn redwood from the White Lake fossil beds, marvelling when we were told they were millions of years old. And the cliffs nearby were made of lava, meaning that there were volcanoes here once, too. Our father had read Hugh Nasmith's late glacial history of the Okanagan, so we learned that the Big Hollow of our toboggan runs was a glacial kettle, that the silt cliffs we clambered over were bottom sediments of Glacial Lake Penticton and that an ice dam at McIntyre Bluff south of Vaseux Lake had dammed the Okanagan drainage, sending the water north instead of south. One of our favourite spots was Coyote Rock, a giant boulder perched atop a tower of unconsolidated rock and silt. How did it get there? Geology is history, and every landscape is full of fascinating geological stories.

The Okanagan has a complex geological history, and in that respect it is a characteristic corner of British Columbia. The original gentle shores of this province have been squeezed, melted, broken, piled and sheared by collisions with offshore terranes, or pieces of the earth's crust. The foreign rocks of the terranes added to the confusion of local rocks, and then glaciers came and piled immense quantities of rubble on top of the whole tableau.

But while all this was going on, animals and plants lived here—on land, in fresh water and in the Pacific. What did it look like back then? And how did the flora and fauna of British Columbia move, adapt and evolve with the changing landscape?

The deciduous dawn redwood was a dominant tree of swampy areas in the B.C. Interior 45 million years ago. These fossil twiglets are from beds in the White Lake area near Penticton.

Our landscape is still changing. Vancouver Island is being squeezed narrower and higher as you read this. Many earthquakes shake the land and sea each year. The climate is warming and glaciers are receding further, for the moment. And trees, flowers, birds, bugs, bears and fish continue to move across the landscape, testing and altering ecosystems as they go. These are the stories told in the following pages. The link among them all is change, for they are stories of transformation.

PART ONE
Origins

IN BRITISH COLUMBIA, it is hard to ignore geology. Most of us may not understand the rocks around us as fully as we would like, but we are well acquainted with them—they stare at us from mountain cliffs and rugged shorelines every day.

If you were to take a flight across the province—say, from Jasper, Alberta, to Port Hardy on Vancouver Island—you would see a wide variety of geological features. First come the sedimentary rocks of the glistening Rockies, rising abruptly above the forested plains. Next, the snowy peaks of the Cariboo Mountains rise up across the Rocky Mountain Trench, their crystalline rocks tortured by the heat and pressure of unimaginable forces beneath the surface of the earth. Now the broad Interior plateau comes into view, with its flat surface of poured lava surrounding the eroded valleys and canyons of the Fraser and Chilcotin Rivers. And finally, the shining white ice of Mount Waddington rises ahead, towering over the Coast Mountains' choppy sea of granite. The face of the province that you have seen—its two great montane belts, separated by the Intermontane plateau and fringed to the west by the Insular belt with its islands and passages—is the cover of a book in which we may read a long and fascinating story.

In more ways than one, geology is the foundation of natural history. Geological formations not only form the physical base of terrestrial life and control the climate around it but also tell the temporal history of nature. Geology tells us how things came to be the way they are. Moving continents, rising and falling mountain barriers, vast volcanic eruptions and continental ice sheets all played an essential role in creating the diversity of life in British Columbia today.

In the field of geology, an oft-quoted maxim is "The present is the key to the past," meaning that in order to understand old rocks, we have to look at geological processes that are at work today. The reverse is also true: to understand the diversity of landscape and life as we see it now, we need to delve into the deep past to see how the land of British Columbia came to be.

The Building of British Columbia: Plate Tectonics

British Columbia is part of the North American Cordillera—the mighty set of mountain ranges that stretch from northern Alaska to southern Mexico (Map 1, overleaf). This mountainous landscape arose through plate tectonic processes. Plate tectonics is how the earth works. Its crust and underlying relatively stiff upper mantle form a carapace of plates like the bones of a baby's skull before they suture and lock together. The plates are constantly moving—some growing, some shrinking—at about the speed a fingernail grows. The key to the Cordillera is a long history of interactions between the western edge of the continent, the plates that make up the floor of the Pacific Ocean, and the small, mobile pieces of crust in between that have been created, that have evolved and that have shifted between ocean and land.

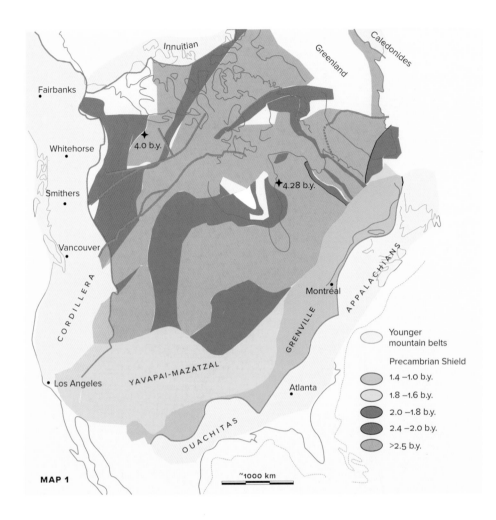

Innuitian

Caledonides

Greenland

Fairbanks

Whitehorse

✦ 4.0 b.y.

Smithers

✦ 4.28 b.y.

Vancouver

APPALACHIANS

Montréal

CORDILLERA

GRENVILLE

Los Angeles

YAVAPAI-MAZATZAL

Atlanta

OUACHITAS

	Younger mountain belts
	Precambrian Shield
	1.4 –1.0 b.y.
	1.8 –1.6 b.y.
	2.0 –1.8 b.y.
	2.4 –2.0 b.y.
	>2.5 b.y.

MAP 1

~1000 km

Our planet is unique in having plate tectonics. The constant swirling and recycling of the ocean's rocky floor requires that the planet's interior—like Goldilocks's bowl of porridge—must be just right, not too hot and not too cold. Plate tectonics results from a balance between subduction—the sinking of oceanic plates at trenches like the modern Cascadia subduction zone off British Columbia's west coast—and spreading at ocean ridges, where new crust is created by the rise of hot material in the earth's mantle, as is happening at the Juan de Fuca Ridge a little farther west (Figure 1, overleaf). The process of plate tectonics as we know it began sometime in Precambrian time. Exactly when is a matter of current discussion. But geologists agree that before then, the young earth was too hot and the plates were too buoyant to sink deep into the mantle. Eventually, the planet will cool to the point that upward rise of mantle and melting of basalt to supply the ridges will fail—but we have billions of years left before that time. Meanwhile, the plates shift constantly, slowly, inexorably, building mountains while we sleep, only rumbling their intentions with earthquakes from time to time.

The modern North American continent is constructed like a chocolate-covered nut (Map 1). At its core is an ancient continent, or craton, called Laurentia. Laurentia holds the record for the oldest rocks yet dated on earth, announced by Quebec researchers in 2008 as 4.28 billion years old. Compared with the craton, the rocks that make up the outer continent margin—including those of the Cordillera—are much younger, generally less than 700 million years. All of them have been added to the original continent. There are piles of sedimentary rock that once lay at its outskirts but rode up over it during periodic collisions. There are also continental fragments that had split from the continent but were later pushed back against it, parts of the margin that were dragged sideways by the motion of offshore oceanic plates. Some of the added pieces are actually crustal wanderers that crossed oceans to reach the western reach of the growing continent, and they play their role in building mountains there.

Not that Laurentia simply sat there, waiting for all this to happen. Its story, too, is that of a wanderer. It has been part of two supercontinents, and probably others before them, in the endless

FIGURE 1: THE PLATE TECTONIC SETTING OFF THE COAST OF BRITISH COLUMBIA TODAY. Magma from the earth's mantle rises upward along the Juan de Fuca Ridge, cooling to form new ocean floor. The walls of the ridge are pulled apart by the same convection currents, and the Pacific Plate and the Juan de Fuca Plate grow symmetrically on either side of the ridge. Where the latter plate encounters westward-moving North America, it slides beneath the continental shelf and descends into the mantle. When it reaches depths of 150 to 100 kilometres, the plate partly melts again, and the resulting magma rises up to reappear as the volcanoes of the Cascade-Garibaldi Arc—among them Mount Meager, Mount Garibaldi, Mount Baker, Mount Rainier and Mount St. Helens. *Adapted from C.J. Yorath, Where Terranes Collide, p. 123.*

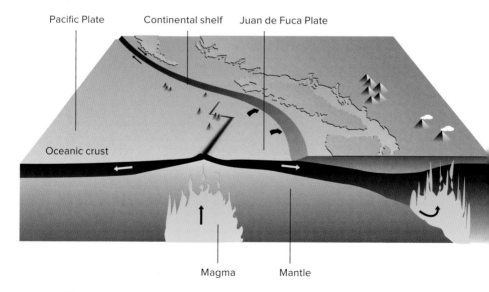

Pacific Plate   Continental shelf   Juan de Fuca Plate

Oceanic crust

Magma   Mantle

flamenco of approach and spurn, touch and turn away that has marked the earth's rocky carapace ever since it formed. The breaking up of the Precambrian supercontinent Rodinia 750 to 550 million years ago did not create our Cordillera—that was many eons later— but it made the Cordillera possible. Without that breakup, Laurentia would have lain serenely within a vast continental interior: a prairie, perhaps, or a vast plain of lakes and wetlands, its smooth, low sur- face unbroken by even a dream of mountains.

But as it happened, towards the end of Precambrian time, a rift formed in what is now southern British Columbia, one of the many that fragmented the world continent Rodinia into many pieces. Whatever was on the other side of that rift—Australia, Antarctica and Siberia each has its advocates—moved slowly and stately away to the west. The Pacific Ocean was born, and the whole tectonic drama of Cordilleran evolution could begin.

The Cordilleran terranes are pieces of once-mobile crust that make up much of the Cordillera, extending west to the Pacific Ocean from an eastern edge in the Omineca Mountains. On Map 2 (page 16) you see them divided into realms, according to their origins. The peri-Laurentian terranes lie between the Omin- eca Mountains and the western Coast Mountains and underlie the Intermontane region in between. They once were the bedrock of arc- shaped chains of volcanic islands and small oceans that lay west of the old continent, in a complex and evolving geography comparable with the other side of the Pacific Ocean basin today. Think of Japan, perhaps, or the Philippines. One of the ancient island arcs is named Quesnellia, after the town of Quesnel. It runs from there north to the Yukon border east of Teslin Lake and south past Princeton. The other old island arc, Stikinia, spans western British Columbia from Bella Coola to Atlin. Island arcs form above subduction zones. Their volcanoes build from lavas and explosive volcanic deposits that orig- inate as melts of the subducted plate as it plunges down into hotter and hotter mantle.

TABLE 1: BRITISH COLUMBIA GEOLOGICAL TIMELINE

Millions of years ago		ERA	PERIOD
570	The diversification of complex life marks the beginning of the Paleozoic Era.	PALEOZOIC	
530	The Burgess Shale fauna lives along the western shelf of North America, near the present site of Field in the western Rockies.		
400	Oldest rocks of Stikinia and Quesnellia formed.		
245	Ninety-six per cent of all marine species become extinct, bringing the Paleozoic Era to an abrupt end.		
200	Stikinia and Quesnellia lie far off the west coast.	MESOZOIC	
181 to 170	Intermontane terranes dock; the Rockies begin to build.		
115	Dinosaur trackways are preserved along the present Peace River.		
120	Western ranges of the Rockies are stacking up.		
175 to 100	Insular terranes dock; renewed subduction after the collision creates the granitic rocks of the present Coast Mountains beneath the new continental margin.		
100	Main ranges of Rockies are forming. Small terranes of the western Cascades and southern Coast Mountains merge, still well south of the British Columbia coast.		
85	Farallon Plate rifts; Kula Plate spreads north, smearing the continental margin to the northwest.		
65	The dinosaurs and many other life forms become extinct, bringing the end of the Mesozoic Era. Tertiary Period begins.	CENOZOIC	TERTIARY
60	Pressure eases and mountain building in the Rockies stops. The crust stretches and thins, and the land to the west of the Rockies breaks away and drops, beginning to create the Southern Rocky Mountain Trench.		
55	Change in plate movement brings Pacific Rim terrane to the southwest coast of Vancouver Island. Crescent terrane arrives sometime later (before 40 million years ago), forming the southwestern tip of the Island.		

TABLE 1: BRITISH COLUMBIA GEOLOGICAL TIMELINE

Millions of years ago		ERA	PERIOD
55	(continued. . .) The compression of the Kula Plate against the North American Plate creates the linear folds and faults of the Gulf Islands.		
50	Relaxation in tectonic pressure opens the Okanagan Valley.		
45	Kula Plate is consumed by subduction underneath the North American Plate, and the compression that had built the original Coast Mountains ceases.		
55 to 36	Lava flows fill many basins in the Interior (e.g., Kamloops Lake area, White Lake in south Okanagan).	CENOZOIC	TERTIARY
10	Original Coast Mountains have eroded to a lower range than today.		
21 to 6.8	Vulcanism is common along a belt from the Coquihalla River through to Pemberton.		
5	Change in the subduction zone of offshore plates gives birth to the Cascade Volcanic Arc and heats and expands crust beneath the Coast Mountains, beginning a 2-kilometre uplift that continues today.		
15 to 2	A multitude of lava flows lays down the plateaus of the Cariboo and Chilcotin.		

Years ago		ERA	PERIOD
2 to 1.6 million	Beginning of the ice ages.	CENOZOIC	QUATERNARY
20,000	Mount Garibaldi is born.		
14,000	Maximum extent of Cordilleran Ice Sheet is reached in the last glaciation in the south.		
13,000	Parts of southwestern coastal lowlands are ice-free.		
10,000	Glacial ice has left all the lowlands.		
7300	Mount Mazama erupts, creating Crater Lake, Oregon, and covering southern British Columbia in volcanic ash.		
250	The last volcanic eruption in British Columbia: lava from a volcanic eruption in the Nass Valley dams the Tseax River and destroys Nisga'a villages.		

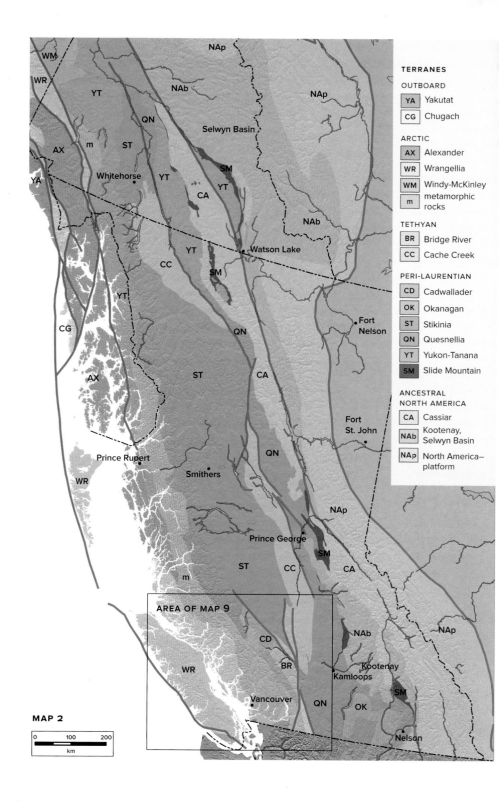

NAp

WM

WR

YT

NAb

NAp

QN

Selwyn Basin

AX   m   ST

SM

YA

Whitehorse   YT   YT

CA

NAb

YT   Watson Lake

CC

SM

YT

NAp

Fort
Nelson

CG

QN

AX   ST   CA

Fort
St. John

Prince Rupert   QN

WR   Smithers

NAp

Prince George

SM

ST   CC   CA

m

AREA OF MAP 9

NAb   NAp

CD

Kootenay
BR

WR   Kamloops

Vancouver   QN   OK   SM

MAP 2

0   100   200

km

Nelson

TERRANES

OUTBOARD

| YA | Yakutat |
| CG | Chugach |

ARCTIC

AX	Alexander
WR	Wrangellia
WM	Windy-McKinley
m	metamorphic rocks

TETHYAN

| BR | Bridge River |
| CC | Cache Creek |

PERI-LAURENTIAN

CD	Cadwallader
OK	Okanagan
ST	Stikinia
QN	Quesnellia
YT	Yukon-Tanana
SM	Slide Mountain

ANCESTRAL
NORTH AMERICA

CA	Cassiar
NAb	Kootenay, Selwyn Basin
NAp	North America– platform

Parts of these volcanic island chains were founded on rifted fragments of Laurentia (to imagine a rifted fragment, think of California west of the San Andreas Fault—this piece of the continent is being pushed inexorably north and will eventually sail past the west coast of British Columbia). The Slide Mountain terrane is the Late Paleozoic seafloor of a minor ocean that grew between one of these rifted chunks and the mother continent. It is spectacularly exposed in the Cassiar Mountains of far northern British Columbia, forming dark peaks of basalt and deep-water sediments where it now rests atop the pearl-grey limestones of western Laurentia. It is as if the floor of today's Sea of Japan, a small ocean that for the last 20 million years or so has been widening between Japan and mainland Asia, were to be shoved back up on top of Korea, and then the whole pile uplifted and carved into mountains.

Compared with the relatively local peri-Laurentian terranes, those of the Tethyan and Arctic realms have travelled astounding distances to arrive in their present Cordilleran berths. Among the many lines of evidence for their exotic origins, fossils are one of the most compelling. The Cache Creek terrane forms a discontinuous strip in the British Columbia Interior, surrounded to the east, north

THE BURGESS SHALE: A REMARKABLE WINDOW INTO THE DISTANT PAST

About 530 million years ago, where today the Rockies tower over the small community of Field, an underwater escarpment snaked along the ocean floor west of the continental shore. Known as the Cathedral Escarpment, it was the wall of a limestone reef built by calcareous algae. At the base of this wall, a thriving community of animals lived on, in and above the muddy sediments that drifted down from above. Periodic mudslides buried these animals in fine silts, which hardened over the millennia to form shale. Beginning about 175 million years ago, tectonic forces pushed the fossil-bearing shale many kilometres east and raised it thousands of metres above sea level, part of the newborn Rocky Mountains.

Today this site is known as the Burgess Shale, named for the pass that is traversed to reach it, and it is the most remarkable fossil bed in the world. Few sites can boast such finely detailed fossils, and few preserve animals of this age. The Burgess fauna lived only geological moments after what has been called the Cambrian explosion, the great diversification of complex animal life, and the Burgess Shale provides the sharpest picture we have of that extraordinary period. It also provides an extensive picture, with over 73,000 specimens of 140 species recovered to date.

Perhaps the most amazing feature of the Burgess fauna is its diversity in anatomical designs for life. Some of the fossils are recognizable ancestors of today's animals, but others are unlike anything in modern oceans. There are oddities such as *Opabinia*, a segmented creature with five eyes and a clawed frontal "nozzle," and *Anomalocaris*, a large (about 50-centimetre-long) predator with stalked eyes, a mouth that looks like a pineapple ring and a series of lobed fins. There is *Hallucigenia*, an odd relative of modern onychophoran worms that has a series of paired spikes down its back, so odd that paleontologists first reconstructed it upside down, walking on stilts. Stephen Jay Gould, in his book *Wonderful Life*, used the term "disparity" to describe this richness of body plans in the Burgess fauna. He and others considered many of these strange creatures to be representatives of previously unknown phyla, now long extinct. However, subsequent research has revealed that only a few of these animals are truly novel; rather, most can be better described as strange (to us) elaborations of the designs found in present-day phyla such as arthropods and annelids.

OPABINIA

ANOMALOCARIS

> Why is Opabinia...not a household name in all domiciles that care about the riddles of existence?
>
> STEPHEN JAY GOULD, *Wonderful Life*

But what happened to the bizarre *Opabinia* and the fearsome *Anomalocaris*? Why aren't their descendants swimming in today's oceans? One could ask the same question regarding the more familiar trilobites and dinosaurs. The textbook answer would be that a few of the Cambrian body plans were more efficient and successful than the others; these body plans prevailed, while others vanished. But if you compared the "failures" with their contemporaries, the early representatives of modern groups, you would find that there is no way to predict that one would prevail and another disappear. At times of great extinctions (as happened at least three times in the Cambrian period), evolutionary change is perhaps more related to historical accidents than to everyday evolutionary "fitness." As Gould puts it, if we rewound the tape of life and let it play again, we would get an entirely different world.

Within the animal community preserved in the Burgess Shale is a rare, small, wormlike creature called *Pikaia gracilens*. Named after nearby Pika Peak, it is believed to be one of the world's first known chordates. Gould sums up his thesis of disparity followed by decimation using this oldest of our chordate relatives as an example:

> And so, if you wish to ask the question of the ages—why do humans exist?— a major part of the answer, touching those aspects of the issue that science can treat at all, must be: because Pikaia survived the Burgess decimation. This response does not cite a single law of nature; it embodies no statement about predictable evolutionary pathways, no calculation of probabilities based on general rules of anatomy or ecology. The survival of Pikaia was a contingency of "just history." I do not think that any "higher" answer can be given, and I cannot imagine that any resolution could be more fascinating.

The public can see this world-class fossil site through the Burgess Shale Geoscience Foundation (www.burgess-shale.bc.ca), which offers guided, day-long hikes during the summer.

HALLUCIGENIA

PIKAIA

and west by more local peri-Laurentian terranes. Its southern exposures can be seen around Cache Creek and Clinton and as far west as the white limestone bluffs of Marble Canyon. On the drive north of Cache Creek on Highway 97, some of the nearby low hills are made of curiously bare, crumbly, dark green to blue-green scree. This is serpentinite, a rock that once made up the deep mantle underpinnings of oceanic crust. Serpentine is a stone that grows little moss, and still less complex forms of vegetation. Compared with continental crust, which has benefited from the distillation of nutrients in generations of magmas and of sedimentary cycles, mantle is a poverty-stricken substrate composed of silica, magnesium, iron, nickel, cobalt and precious little else. Few plants can survive in its nutrient-poor soils. But its presence here delights the geologist, because its exhumation from deep mantle to grassland demonstrates a powerful process of planet-scale plate motion and, more specifically, a dramatic collision of an oceanic plate with the continent.

If you were to look closely at the limestones around Marble Canyon you would find, along with corals, some unassuming little fossils that look like fat grains of wheat. They are fusulinids, a now-extinct family of foraminifera (shelled amoeboid organisms) that flourished in warm Late Paleozoic seas. The youngest Marble

YABEINA

Canyon fusulinids are Late Permian, and some are of the genus *Yabeina*. These small foreign creatures have no known relatives in or near Laurentia, but they and all their cousins can be found in their billions in the Permian limestones of China and Japan. In Permian time, long before the continental collisions that drove the Alps and Himalayas skyward, a bend of ocean called the Tethys lay surrounded by Europe, Siberia, Africa, India and Antarctica, with the continental fragments that now make up China on its east. *Yabeina* grew prolifically there.

The Marble Canyon limestones are thought to have been reefs built on an ocean island somewhere on that

side of the Pacific. After that, the island must have become entrained in an eastward-moving oceanic plate, reeled towards the Laurentian margin by rapid subduction under its fringing island arcs, Stikinia and Quesnellia.

Outside and west of the peri-Laurentian terranes in British Columbia lie the Insular terranes, Wrangellia and the Alexander terrane—the bedrock of Vancouver Island, Haida Gwaii (the Queen Charlotte Islands) and the islands of the Inside Passage. These rocks are also exotic but probably with an entirely different origin than that of the Cache Creek terrane: they once were part of the Arctic realm. Their older parts formed and evolved somewhere near northern Scandinavia and eastern Siberia until in mid-Paleozoic time, when they were propelled westward through the Arctic Ocean and

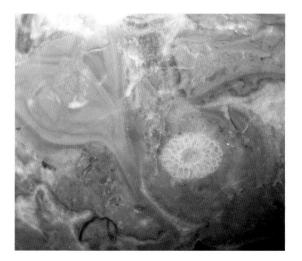

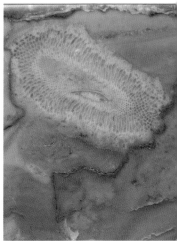

Silurian sphinctozoan sponges, from Alaska Prince of Wales Island—natives of the Ural Mountains region, brought to these distant shores by plate tectonics.

into the Pacific. Again, some of the key evidence is fossils. For instance, some unusual early Paleozoic sponges (480 to 420 million years ago) are found in the Alexander terrane on Prince of Wales Island in southeastern Alaska just north of Prince Rupert. Other than the Alexander terrane, these particular sponges are found only in terranes of northwestern Alaska and Oregon, and in the southern Ural Mountains.

The transport of the Arctic and Insular crustal fragments westward across the Arctic seaway left its traces in glancing mid-Paleozoic collisions recorded in the rocks of the Canadian Arctic Islands and the Brooks Range of northern Alaska. Unlike the ocean floor that ferried the Cache Creek oceanic islands towards the Laurentian margin under traction from its subduction zones, the Arctic terranes were fragments of volcanic island arc and continental origin that might have transited between northern Laurentia and Siberia by a mechanism like the recent history of the Caribbean ocean (Maps 3–8). In the Caribbean, an island arc that once lay next to the Pacific Ocean reformed into a giant, bulging loop that surged over a thousand kilometres across to the Atlantic side, its ends colliding with the Bahama Banks to the north and Venezuela to the south. This incredible journey is well documented by geological observations. It has taken about 60 million years to accomplish, and is still happening, with the eastward migration of the Lesser Antilles island chain. The tragic earthquake in Haiti in 2010 was a catastrophic release of pent-up strain on the Enriquillo-Plaintain Fault, one of the great faults that separates the eastward-moving Caribbean plate from westward-moving North America.

The "loopiness" of island arc chains in general—think of the graceful festoons of the Aleutians, Kuriles and Marianas around the north and west of the Pacific—is caused by the oceanward advance of island arcs towards their subduction zones. The shorter the total length of the arc, the faster its advance because the easier it is for mantle to flow around its ends and into the gap behind it, where a new little ocean opens wider with time. Short arcs clock high rates of forward migration—1.8 centimetres a year for the Lesser Antilles, 5.7 centimetres a year for the Scotia arc southeast of Tierra del Fuego and 6.8 centimetres a year for the Calabrian arc, a tiny obscure feature of the Mediterranean Sea. By contrast, the centre of the 4000- kilometre-long Andean arc is thought to be actually retreating at 0.7 centimetres per year. With this in mind, it is easy to imagine that the short arc segment between Laurentia and Siberia would have been a prime bet as a fast forward traveller.

The evolution of marine faunas in the Insular terranes attests to the terranes' westward migration. By Late Paleozoic time, instead of eastern Arctic forms, fossils in them are typical of northern Pacific waters. They were not yet interacting directly with anything on the western Laurentian margin, but they were getting close enough to play their part in the events to come.

Collisions and Upheavals: the Continent Grows West

The mid-Jurassic, about 185 to 170 million years ago, was a time of crisis and profound change in the Cordillera. Before then, the peri-Laurentian terranes formed a dynamic, shape-shifting zone west of Laurentia. Farther west, the Insular terranes shifted and rifted, still all on their own. After the mid-Jurassic, all of these massive crustal blocks came together to collide and coalesce, heave and pile, thrust and thicken, creating the Cordilleran mountains that we know now.

What happened?

The key is in the timing. The Insular terranes collided with the outer margin of the peri-Laurentian terranes, in what is now the western Coast Mountains, in the mid-Jurassic. In southeastern British Columbia, in the Goat Range near New Denver, the peri-Laurentian terranes were first thrust up on the sedimentary apron of the continent—in the mid-Jurassic. The youngest ocean-bottom deposits in the Cache Creek terrane that represent the end of the terrane's existence as an open ocean are from the late Early Jurassic. The volcanoes of Stikinia and Quesnellia all shut down in the

MAPS 3 TO 8. THE TECTONIC EVOLUTION OF WESTERN NORTH AMERICA (*pages 25–29*). The assembling of the west coast of North America is a complex story, and this series of maps serves as a visual guide to the wanderings of terranes. An approximate outline of the present continent is in blue, and the inferred extent of continent through time is shown by grey shading. Orange shading shows active mountain belts.

MAP 3. SILURIAN (425 million years ago) (*facing top*). The Arctic terranes (yellow) that now occupy coastal B.C. and part of Alaska probably originated near the northern end of the Caledonian mountain belt, between the continents of Laurentia, Siberia and Baltica.

MAP 4. EARLY DEVONIAN (395 million years ago) (*facing bottom*). The westward travel of the Arctic terranes towards Panthalassa (the precursor to the Pacific Ocean) is believed to have been propelled by a Caribbean-style subduction zone. This subduction zone, its island arcs and continental fragments travelled rapidly along a Paleozoic Northwest Passage.

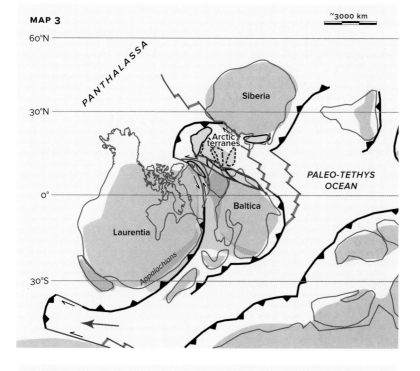

MAP 3

~3000 km

60°N

PANTHALASSA

Siberia

30°N

Arctic terranes

PALEO-TETHYS OCEAN

0°

Baltica

Laurentia

Appalachians

30°S

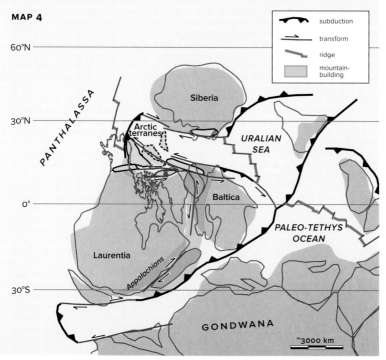

MAP 4

	subduction
	transform
	ridge
	mountain-building

60°N

PANTHALASSA

Siberia

30°N

Arctic terranes

URALIAN SEA

0°

Baltica

PALEO-TETHYS OCEAN

Laurentia

Appalachians

30°S

GONDWANA

~3000 km

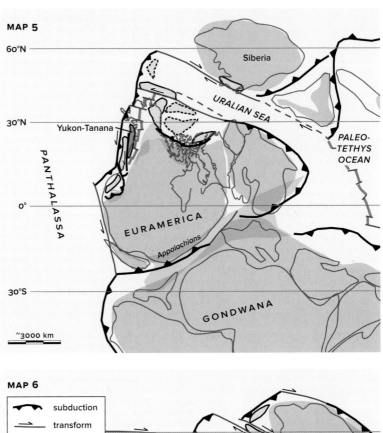

MAP 5

60°N

Siberia

URALIAN SEA

30°N

Yukon-Tanana

PALEO-
TETHYS
OCEAN

P
A
N
T
H
A
L
A
S
S
A

0°

EURAMERICA

Appalachians

30°S

GONDWANA

~3000 km

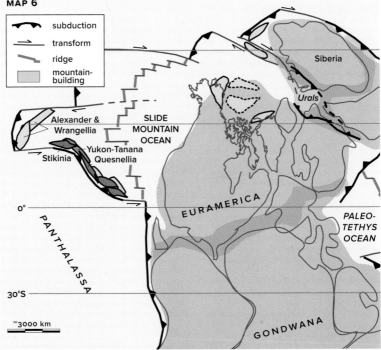

MAP 6

subduction
transform
ridge
mountain-
building

Siberia

Urals

Alexander &
Wrangellia

SLIDE
MOUNTAIN
OCEAN

Yukon-Tanana
Quesnellia

Stikinia

PALEO-
TETHYS
OCEAN

0°

EURAMERICA

P
A
N
T
H
A
L
A
S
S
A

30°S

GONDWANA

~3000 km

mid-Jurassic, signifying the death of the subduction zones that had fed them. Whatever triggered these sweeping and simultaneous changes must have been at a scale vaster than all the terranes taken together.

The likely cause lies in global plate tectonics. In Middle to Late Paleozoic time, Laurentia had become incorporated into the super-continent Pangaea, by collisions with Europe and South America that built the Appalachians. But supercontinents, like empires, carry the seeds of their own demise. Like Rodinia before it, Pangaea began to break up in the Early Jurassic. The North Atlantic began to open about 180 million years ago—first a crack, then a seaway, and then, by the mid-Jurassic, a nascent ocean. A new continent, North America, with old Laurentia in its core, started to move ponderously westward. The once-independent terranes of the Cordillera simply got in the way.

MAP 5. LATE DEVONIAN–MISSISSIPPIAN (360 million years ago) (*facing top*). The northward shift of Euramerica (the now-combined Laurentia and Baltica), during its collision with Gondwana, results in the formation of a subduction zone along the west coast. This subduction begins from the small Caribbean-type zone and moves some of the Arctic terranes southward. The hot upwelling beneath the subduction zone causes a rift in Laurentia, giving birth to the first of the peri-Laurentian arc terranes along the west coast, including the Yukon-Tanana terrane (blue-green).

MAP 6. PENNSYLVANIAN–EARLY PERMIAN (300 to 285 million years ago) (*facing bottom*). Westward retreat of the subduction zone leads to widespread island arc volcanism (green) that develops on fragments of western Laurentia (blue-green) and on some of the Arctic terranes (yellow) that had recently arrived. These new arc terranes include the early expressions of Stikinia and Quesnellia. The Slide Mountain Ocean develops between the island arc and the continental margin in the wake of the westward migration of the arc, mirroring what is happening in the present-day Sea of Japan. The Insular terranes of coastal B.C. (Alexander and Wrangellia) come together far out in Panthalassa. Alexander is one of the Arctic terranes with northern European origins.

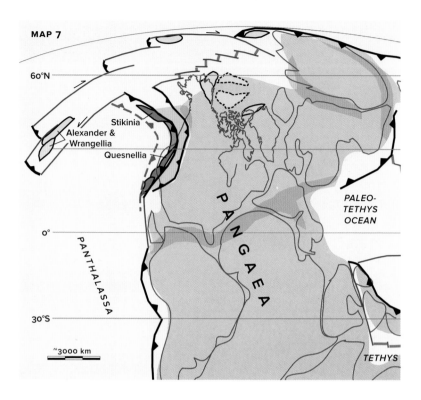

MAP 7

60°N

Stikinia
Alexander &
Wrangellia
Quesnellia

PALEO-
TETHYS
OCEAN

PANGAEA

0°

PANTHALASSA

30°S

~3000 km

TETHYS

MAP 7. LATE PERMIAN—EARLY TRIASSIC (250 million years ago) (*above*). All major continental masses have converged to form the supercontinent Pangaea. Along its west coast, subduction has reversed to consume the Slide Mountain Ocean, returning the peri-Laurentian arc terranes (green) to near the continental margin. Later in Triassic time, subduction flips once more (dashed grey line), and arc volcanism flourishes again on Quesnellia and Stikinia. Alexander and Wrangellia remain at large in Panthalassa.

MAP 8. EARLY JURASSIC (190 million years ago) (*facing*). The Atlantic Ocean is born, growing in the rift between North America, Africa and part of Europe, and propelling North America westward. Buckling of the peri-Laurentian (Intermontane) terranes traps part of the ancient Pacific Ocean floor that was brought to North America by the subduction conveyor belt from far reaches—the Cache Creek terrane. At the same time, the westward-moving North America is on a collision course with the Insular terranes lying offshore in the Pacific. The ultimate collision will build the mountains of western North America and shape the final terrane patchwork (Map 1, page 10).

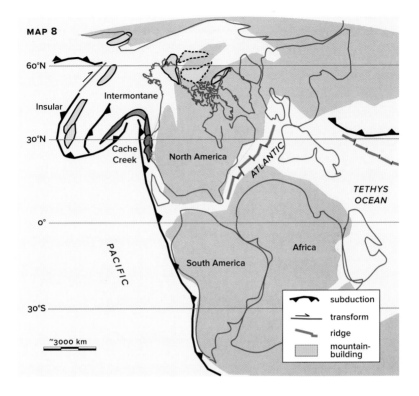

MAP 8

The result was our mountains—low ones at first, with the initial collisions, but as the continent continued its inexorable course, sedimentary strata at its margin piled up like snow in front of a vast, majestic snowplow, riding up and over eastward to make the shingled stack that later would be sculpted into the modern Rockies. The physiography of British Columbia—its twin backbones of the Coast Mountains and the Ominecas and Rockies separated by the more subdued Intermontane belt—is the result of the two, slow-motion, simultaneous collisions. Where the Intermontane terranes piled up onto the old continental margin, the Omineca and Rocky Mountains rose. Where the Insular terranes collided with the outer edge of the Intermontane terranes, the Coast Mountains were born.

Between 250 and 210 million years ago, before the Atlantic Ocean had opened, the Americas were still fused to Eurasia and Africa in the supercontinent Pangaea. Dinosaurs had yet to appear on land, but in the deep waters off Pangaea's west coast a diverse community of invertebrates, fishes and seagoing reptiles existed. After they died, the abundant, bony fish sank to the bottom and settled in the soft muds and silts. Many were preserved as fossils as the soft ooze was slowly compressed into

shale and mudstone by later sediments. Millions of years later, the Intermontane terranes collided with this oceanic shelf, now the western continental shelf of North America, and its deep sedimentary rocks were squeezed, broken and piled up on one another to form the Rocky Mountains. Today one of the best sites for fossils of Triassic fish is above Fossil Fish Lake, near Wapiti Lake, northeast of Prince George.

The bony fishes of the Triassic can be divided into the lobe-finned and ray-finned fishes. The lobe-finned fishes, which are represented among living fishes only by

THE TRIASSIC COELACANTH
WHITEIA (50 CM LONG).

THE RAY-FINNED FISH BOBASTRANIA
(A SMALL SPECIMEN, 30 CM LONG).

The Omineca–Rocky Mountain Collision Zone

As North America drove under its western neighbours during the Middle Jurassic, large pieces of Quesnellia and the Slide Mountain terrane began to peel off the oceanic plate. Some slices up to 25 kilometres thick overrode the continental margin, becoming stacked like pancakes on top of it. This stacking makes it difficult to say precisely where the old edge of North America lies today. The rocks of the terranes and the old continental shelf were squeezed and folded to form the Columbia, Omineca and Cassiar Mountains. In some

the coelacanths and lungfishes, would eventually give rise to the amphibians and thus ultimately the other vertebrate classes. This group is represented in the Wapiti Lake fossils by *Whiteia,* a coelacanth similar to those found today off Africa's eastern shores. *Whiteia* is thought to have been a stalking or lunging predator. Its numerous, flexible fins are well placed for proficient locomotory control, but *Whiteia* wouldn't have been a fast swimmer.

The ray-finned fishes would eventually give rise to the other fishes that swim on earth today. One primitive group of

THE RAY-FINNED FISH ALBERTONIA
(35 CM LONG).

ray-finned fishes is directly related to the modern sturgeons and paddlefishes. This group is represented at Wapiti Lake by several genera, including *Bobastrania. Bobastrania* is a large fish (up to 1 metre long) with crushing teeth that probably allowed it to eat hard-shelled animals such as lobsters or shrimp.

Another group, represented today only by the bowfin, is believed to be on the direct evolutionary line to the teleosts, the group that makes up the bulk of the modern fish fauna. The bowfin's group is represented at Wapiti by *Albertonia,* a large, strong-swimming fish with small teeth, suggesting that it fed on plant material or small prey. Its pectoral fins are large, like those of modern flying fish, but this fish's shape makes it an unlikely flyer. The pectoral fins were more likely used for manoeuvring, for detecting or even tasting potential prey, or perhaps for sexual displays.

areas the intense compression and consequent heating recrystallized the rocks into the metamorphic rocks of the Omineca and Monashee Mountains and the Quesnel and Shuswap Highlands. Partial melting in some regions gave rise to local intrusive igneous rocks.

Compression continued, and the thick layers of sedimentary rocks covering the continental core were pushed ever eastward in front of the colliding wedge and were squeezed, folded and

pages 30–31: A river of golden volcanic rock flows down a steep slope in the Ilgachuz Mountains, north of Anahim Lake.

telescoped (Figure 2). The sedimentary layers first were deformed into waves like those in a carpet being pushed. But the strong, resistant limestone layers broke when folded and became stacked up one on top of another in gently sloping piles. These breaks are called thrust faults, and the blocks of rocks above the break are called thrust sheets. By 120 million years ago, the western ranges of the Rockies were stacking up. A deep depression, the Rocky Mountain Trough (not Trench) formed east of the mountain-building wave, the result of the tremendous weight building up on the edge of the continent. The rapid uplift caused massive erosion of the new mountains, and sediments soon piled up in the trough's inland sea, forming thick deposits of mudstone and shale.

FIGURE 2: THE FORMATION OF THRUST FAULTS AND THRUST-FAULTED MOUNTAINS. *Stage 1:* Compression from the left bends and finally breaks the rock layers. *Stage 2:* The upper sheet of rocks, the "thrust sheet," is pushed over the lower sheet. *Stage 3:* The face of the mountain after erosion. Some of the ancient limestones at the bottom of the sedimentary pile (e.g., layer D) end up on top of the younger shales (e.g., layer B). *Adapted from C.J. Yorath, Where Terranes Collide, p. 9.*

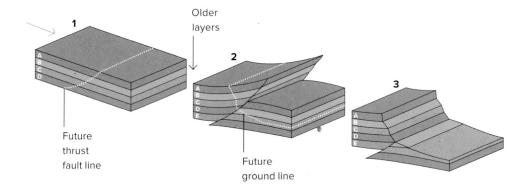

Older layers

1

Future thrust fault line

2

Future ground line

3

The mountain-building wave in the Rockies continued to move eastward. The main ranges were rising about 100 million years ago and, by the time the pushing stopped about 60 million years ago, the eastern

Treadmill Ridge, looking south along the Continental Divide between Jasper National Park and Mount Robson Provincial Park. These gently sloping mountains end in abrupt cliffs to the east, which mark today's eroded edge of a thrust sheet.

ranges and foothills had been created. When all was said and done, the thrust sheets (Figure 2) had been telescoped and shoved up to 250 kilometres eastward from their original position—the rocks of Mount Rundle, at Banff, were originally laid down somewhere around Revelstoke. As the thrust sheets moved to the east and stacked on top of one another, the Rocky Mountain Trough moved eastward ahead of them. But the thrust sheets overtook the shales that had been deposited in the trough's earlier position, and layers of these soft shales were caught between the sheets. The shales erode much more easily than the resistant limestones, and this difference

TRACKING DINOSAURS IN
ROCKY MOUNTAIN FORESTS

The climate was warm 115 million years ago, and dinosaurs were at their zenith. The Coast Mountains had yet to rise, so the Interior was moist, not lying in the rain shadow of a mountain range, as much of it does today. Humid lowland forests of ginkgos and conifers flourished, underlain by stocky cycads and carpets of ferns and seed ferns. The seed ferns were not related to ferns but rather to the seed-bearing cycads. Although much of the vegetation would have appeared strange to us, the conifers would have looked familiar, since they were dominated by early relatives of the redwoods. Pines, although present, were rare, at least in the warm valleys where their remains would become fossils. The record of these forests is preserved today in the great coal fields of the Rocky Mountains—both in the Crowsnest Pass and the Tumbler Ridge–Hudson's Hope region.

Along rivers thick with sediments from the young Rocky Mountains, dinosaurs roamed through the river deltas in the area that is now the east arm of Williston Lake, west of Hudson's Hope. As they strolled along the calm oxbow ponds beside the rivers, they left footprints in the undisturbed mud. Soon floodwaters brought more silt that covered the tracks. For 10,000 to 15,000 years, about 4 centimetres of sediments were laid down every year, and many generations of dinosaurs left signs of their daily activities in layer after layer of mud, silt and sand. These layers, buried under the growing Rocky Mountains, were eventually compressed into sedimentary rocks.

Millions of years later, the Peace River cut through these layers of rock and revealed the tracks of its long-vanished fauna. Although first reported in 1922, the dinosaur trackways were not studied intensively until the site was about to be flooded by the W.A.C. Bennett Dam. Scientists, working quickly over four summers,

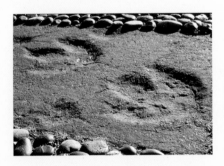

Cast of duck-billed dinosaur *(Ambyldactylus gethingi)* tracks, Royal British Columbia Museum.

Cycad (*Ptilophyllum* sp.) fossil from the Tumbler Ridge coal fields (110 million years old).

located over seventeen hundred tracks, took casts of about two hundred and collected ninety individual footprints. They discovered that the tracks record the existence of at least seven species of dinosaurs and, even more fascinating, tracks of an ancient shorebird and one of the earliest marsupial mammals.

With only a little imagination, we can visualize the stories told by some of the trackways. For example, tracks show that a herd of herbivorous dinosaurs walking south along the riverbank spotted a pack of carnivorous dinosaurs. After quickly turning to the east, they were followed by the carnivores. We don't know the end of the story, since the tracks disappear.

Today the tracks lie deep under the waters of Williston Lake. Once again, sediments from the Rockies gather slowly on their features, perhaps preserving them for a future civilization to discover.

The coal fields of British Columbia also record another fascinating story in the evolution of modern life—the diversification and hegemony of flowering plants. Flowering plants evolved from a branch of the seed fern line and first appear in fossils early in the Cretaceous System—that is, about 140 million years ago. Sixty million years later—30 million years after dinosaurs were wandering the Peace River country—flowering plants now dominated the forests of British Columbia. By then, the Insular terranes had docked, and these forests of broad-leaved trees are recorded in the coal deposits in the Nanaimo area. Although the tree species have been classified by paleontologists in now extinct genera and even families, these forests would have appeared familiar to us, since they superficially resembled modern forests of broad-leaved trees such as oak, poplar, alder, fig, breadfruit, maple and ash.

The early conifers *Elatides curvifolia* (leafy twigs) and *Athrotaxites berryi* (twigs with scale-like leaves), from the Tumbler Ridge coal fields (110 million years old).

The flowering plant *Cupanites crenularis* had pinnately compound leaves. The broad leaf is from an unidentified species. The fossil is from the Comox Formation (80 million years old).

results in a pattern that is seen over and over again in the Rockies—hard, limestone cliffs towering over soft, shale-bottomed valleys (Figure 2, page 34).

The Coast Mountains Collision Zone

As the Insular terranes ran into the Intermontane terranes, a new subduction zone formed to the west near the present continental margin, and a new belt of continental magmatism replaced the old island arcs along the line of the present Coast Mountains. Many separate but coalescing igneous intrusions rose up in a succession of pulses from 170 to 50 million years ago, creating the Coast Mountains batholith, one of the largest bodies of granite and granitoid rocks on the planet.

The heat of all that intrusion softened and weakened the earth's crust and created a second, outboard zone of crustal thickening between

Marble Peak on the mainland, east of Princess Royal Island. The Coast Shear Zone passes through here, as shown by the highly sheared metamorphic rocks of the main peak.

GRANITE AND BASALT:
TWO ROCKS FROM THE SAME MAGMA

All igneous rock is born deep beneath the earth's surface. At a depth of 65 kilometres, the temperature is close to 1200°C, but because of the tremendous pressure, the rock does not melt. When a crack above releases some of the pressure, however, part of the rock liquefies and the pressurized magma moves up towards the surface, cools and solidifies to form new rock.

If the rock cools well beneath the surface, it cools slowly, allowing coarse crystals to grow during the solidification process—crystals such as light-coloured quartz and feldspar and black hornblende or mica. These crystals give the rock a salt-and-pepper appearance. This is granitic rock—whether it is granite, granodiorite, quartz diorite or some other related rock depends upon the relative proportion of its component minerals. True granite is often pinkish because of its high content of potassium feldspar; quartz diorite is more black-and-white.

If the magma reaches the earth's surface, it cools quickly, too quickly for coarse crystals to form. This rock is usually dense and gritty—like basalt. The fast cooling of a basalt lava flow often causes distinctive, angular columns to form perpendicular to the cooling surface.

Like granitic rocks, lavas are classified by their chemical nature—primarily by the amount of silica they contain. Basalts are very fluid lavas and are low in silica, containing about 46 to 51 per cent of this mineral. Andesites, which are the common lavas of the Cascade volcanoes, contain about 54 per cent silica. Mount Garibaldi is a bit unusual, being built of dacite, which is even richer in silica and decidedly more viscous when molten. Rhyolites contain more than 73 per cent silica—probably the most distinctive rhyolite is obsidian, or black volcanic glass.

the advancing new continental margin and the subduction zone. If you drive from Terrace west along the Skeena River towards Prince Rupert, you can see the gneisses that make up the roots of these mountains. Some of them have experienced conditions of pressure and temperature that could only have occurred at 25 kilometres below the surface, showing the amount of uplift that has made, and made again and over again, this maritime mountain range over the years.

The huge volume of granites in the Coast Mountains, as well as the intensity of deformation at later stages in their geological evolution, has made the early history of this mountain range particularly hard to decipher. Crustal thickening and metamorphism 100 to 80 million years ago produced such profound changes that evidence for the initial collision between the Insular and Intermontane terranes has been nearly wiped off the record. Another key structural event that was mostly overwritten by the frenzied later Cretaceous is a series of older faults that can help us understand the geological puzzles posed by the southernmost Coast Mountains and North Cascades.

Terranes of the Southern Coastal Belt

The southern Coast Mountains and the North Cascades, so accessible to hikers and skiers from Vancouver, actually contain some of the most perplexing geology to be found anywhere in the province. Instead of a few big terranes, they are made up of a whole structural stack of little ones (Map 9). Some of these, the Bridge River, Methow and Cadwallader terranes, represent an ocean like the Cache Creek, except that instead of closing in mid-Jurassic time, it did no such thing until halfway through the Cretaceous. Other terranes, like the Chilliwack and Harrison Lake, resemble parts of Stikinia. Then there are piles of little terranes in northern Washington State that represent nothing else in British Columbia and in fact have no known equivalents north of the Klamath Mountains of California. A satisfying solution to this puzzle is finally emerging, thanks to Jim Monger and his colleagues. They point out that the late closure of the Bridge River ocean means that, somehow, Stikinia and the Insular terrane were not even there until about 100 million years ago, unlike farther north where they were well in place 70 million years earlier. Also, the stack of little terranes in Washington were thrust up from the south—neither from the northeast nor southwest, as is the usual case in the main thrust belts of the Coast Mountains or Rockies. These geological anomalies can be explained if

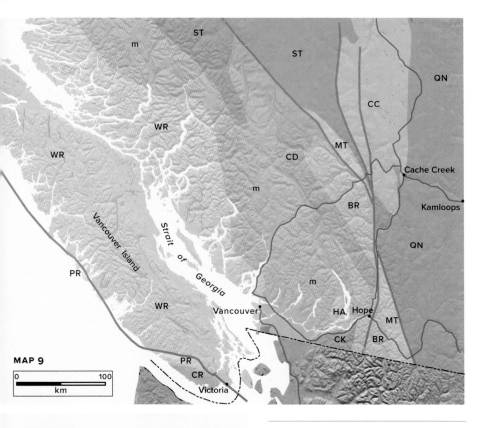

MAP 9

0	100
km	

TERRANES

OUTBOARD

PR	Pacific Rim
CR	Crescent

ARCTIC

WR	Wrangellia
m	metamorphic rocks

TETHYAN

BR	Bridge River
CC	Cache Creek

PERI-LAURENTIAN

MT	Methow
CD	Cadwallader
HA	Harrison Lake
CK	Chilliwack
ST	Stikinia
QN	Quesnellia

MAP 9. TERRANES OF THE SOUTHERN COAST. A detailed look at the terranes of the Cascade Range, southern Coast Mountains and southern Vancouver Island. For the location of this map, see Map 2, page 16.

you imagine that the outer part of the Coast Mountains, along with the Insular belt, moved southward between mid-Jurassic and mid-Cretaceous time, closing off the Bridge River ocean as it went, and eventually rammed into the western Klamaths of northern California. The faults that this happened along have only recently been found. One of them lies under Grenville Channel, that long, narrow straight stretch of water that marks the Inside Passage south of Prince Rupert.

Nowadays, we take for granted northward motion of the Pacific plate relative to North America. This movement is what gives us great modern faults like the San Andreas and Denali, and older ones like the Tintina, the Fraser and the Cassiar. But oceanic plates are fickle and evanescent compared with the long-term existence of continents. It seems that in Jurassic up to mid-Cretaceous time, some plate was out there, charging south with respect to North America and dragging the outer part of British Columbia along with it. It only vanished about 100 million years ago, and other north-travelling plates coupled with the Cordilleran margin and dragged the outer parts of it back up—some might say—where it belongs.

Slipping and Sliding

About 85 million years ago, the Farallon Plate under the Pacific Ocean rifted in two (Figure 3). The northern plate, named the Kula Plate, began spreading in a much more northerly direction than before. Because the North American Plate was still moving west, the new continental margin was now not only squeezed and foreshortened but smeared to the northwest. The crust had to give, and it slid north along faults such as the Northern Rocky Mountain Trench and the Fraser and Queen Charlotte–Fairweather Faults. Along the Northern Rocky Mountain Trench, the land to the west moved certainly 450 kilometres, and possibly up to 750 kilometres northward relative to the Rockies to the east. Faults that separate laterally

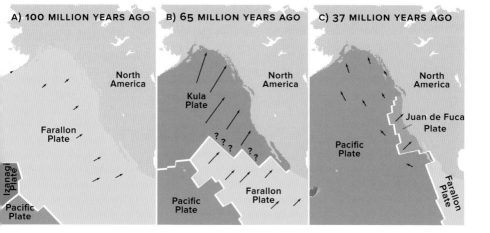

A) 100 MILLION YEARS AGO **B) 65 MILLION YEARS AGO** **C) 37 MILLION YEARS AGO**

North America

Kula Plate

Farallon Plate

Izanagi Plate

Pacific Plate

North America

Pacific Plate

Farallon Plate

North America

Juan de Fuca Plate

Pacific Plate

Farallon Plate

FIGURE 3: SPECULATED PLATE HISTORY IN THE PACIFIC OCEAN. Successive plates are born, grow and are then consumed by succeeding plates. North America is presented as a fixed entity to give a stable reference point, and arrows give relative directions of oceanic plate movement. The lengths of the arrows are proportional to the plates' velocities. In (A), the Farallon Plate dominates the floor of the eastern Pacific Ocean 100 million years ago. At 65 million years ago (B), the Farallon Plate has rifted in two, creating the Kula Plate to the north. By 37 million years ago (C), the Pacific Plate dominates the ocean floor; the Kula Plate has gone and the Farallon Plate is fragmented, creating the small northern Juan de Fuca Plate. *Adapted from H. Gabrielse and C.J. Yorath, eds.,* Geology of the Cordilleran Orogen in Canada, *Fig. 3.3.*

moving surfaces are called strike-slip faults; the San Andreas Fault in California is probably the best known example of such a fault. The resulting pattern from all this faulting and sliding is one of elongate, northwestward-trending terranes, as shown in Map 2, page 16. But the strike-slip faults do not necessarily mark the edges of foreign terranes—the Northern Rocky Mountain Trench, for example, is 50 to 100 kilometres east of the continental margin. The land displaced to the west of it, although originally part of North America, is called the Cassiar terrane.

This squeezing and slipping along the coast of North America continues today—Baja California and all of California west of the San Andreas Fault are sliding slowly northward and will probably

collide with Alaska in 50 million years or so. Off the British Columbia coast, the Queen Charlotte–Fairweather Fault separates similarly sliding chunks of crust.

Relaxation

By 60 million years ago, the Rocky Mountains were a wide band
of magnificent high plateaus and towering mountains probably
over 4000 metres in elevation. But then the pushing stopped. The
Kula Plate found a new subduction route beneath the new continental margin and the tectonic pressure eased. The compressed
crust relaxed and pieces of it began to slide off the thick pile. Along
the western wall of the Rockies from the Robson Valley south, the
Southern Rocky Mountain Trench formed. There, the crust foun-
dered and the western block fell as much as 1000 metres relative to
the mountains on the east. This same faulting process has created
valleys such as the Elk, Flathead and Okanagan.

In the south Okanagan, beginning about 50 million years ago,
a large piece of Quesnellia slid off about 90 kilometres to the west,

facing page: The Southern Rocky Mountain Trench—seen here at Columbia Lake—is a big crack in the earth's crust formed as the crust stretched after the compression of continent-terrane collisions ended.

above: The Okanagan Valley, known as a place to relax, was formed by crustal relaxation. These cliffs of granitic gneiss beside Vaseux Lake are part of the basement of the original continental margin that was exposed when Quesnellia slid off to the west after tectonic pressures were relaxed about 50 million years ago.

exposing the basement rocks of the old continental margin. These ancient rocks can be seen along the east side of Skaha Lake, on both sides of Vaseux Lake and, most spectacularly, in the vertical wall of McIntyre Bluff, just south of Vaseux Lake.

The Latest Collisions

About 55 million years ago, the relative direction of the Kula Plate changed, and it began to slide more northward along a fault similar to today's San Andreas Fault in California. Since the Olympic Peninsula did not yet exist, Vancouver Island projected out into the Pacific and formed a trap for northwesterly slipping terranes. During this period, two terranes were brought up the coast on the Kula Plate and pushed into southern Vancouver Island. The Pacific Rim terrane, which arrived about 55 million years ago, consists of sedimentary and volcanic rocks that now form the southwest coast of the island (Map 9, page 41). After it made contact, the fault between the Kula Plate and North America jammed and a new fault formed slightly farther out in the Pacific. The Crescent terrane, made up of former marine volcanoes that had formed along the fault, was brought alongside sometime later, but before 40 million years ago. In British Columbia, the Crescent terrane's oceanic lavas form the rocks of Victoria's Western Communities; to the south, they make up the Coast Mountains of Washington and Oregon (the rocks of the Olympic Mountains came later). The Crescent terrane is separated from the Pacific Rim terrane by a fault along Loss Creek. The force of this collision did not create big mountain ranges in British Columbia, but it did fold and thrust-fault the sedimentary rocks along the Strait of Georgia to form the ridges and bays of the Gulf Islands.

Over the millennia, the westward-moving North American Plate steadily consumed the Farallon and Kula Plates. About 40 million years ago, the Kula Plate disappeared and any compression between it and North America ceased. Today only a remnant of the once great Farallon Plate remains—the small Juan de Fuca Plate is formed by the emergence of lava along oceanic ridges west of Vancouver Island (Figure 1, page 12). The crust is pulled eastward by the underlying convection currents and is then subducted beneath the continental shelf of British Columbia.

But one terrane collision is occurring even as you read this. In the St. Elias Mountains, where British Columbia, Yukon and Alaska meet, the Yakutat terrane is crunching into the Chugach terrane. There, some of North America's highest and most spectacular mountains rise virtually from the seacoast, and the force of the impact continues to push them up at the remarkable rate of 4 centimetres a year.

Volcanic Belts

One of the most obvious results of plate movements is volcanic activity, and British Columbia has had its share of eruptions ever since Pacific plates began actively subducting beneath the continental margin. Many of the terranes that were added to the British Columbia coast brought substantial volcanic material with them, but much has happened since the collisions as well.

As compression from colliding terranes ended and the crust relaxed and thinned or was pulled northward by tectonic forces, a period of intense volcanic eruptions began. From about 55 to 36 million years ago, lava flows filled deep basins all the way from Yellowstone to the central Yukon. Examples of these rocks can be seen along the northern shore of Kamloops Lake, where 1450 metres of volcanic sediments and lava are exposed in the Tranquille area, and in the south Okanagan, where 2400 metres of sedimentary and volcanic deposits are laid down in the White Lake basin.

From about 21 to 6.8 million years ago, volcanism was common along the Pemberton Volcanic Belt from the Coquihalla valley through to Pemberton. Behind this belt of volcanoes, there was

THE EOCENE SCENE

About to 55 to 50 million years ago, as the Eocene Epoch dawned, British Columbia was warmer than it is today. The last dinosaur had died 10 million years earlier, and the earth's climate was beginning to cool after its tropical days during the Mesozoic Era. On the southwest coast, the Pacific Rim terrane was colliding with Vancouver Island, and the Crescent terrane wasn't far behind.

Vulcanism was widespread in the Interior, as tectonic movement squeezed the western terranes north relative to North America. Ash-rich sediments gathered in the abundant swamps and lakes of the Interior and preserved many fine fossils, giving us a rare glimpse of upland environments of that time. In British Columbia, significant Eocene fossil sites are known in the Smithers area, the Quesnel area, Horsefly, the Thompson River basin, the Nicola Valley, the Princeton area and the Okanagan Valley.

At moderate elevations, the forests were diverse, far richer in species than today's forests. They were dominated by deciduous conifers, holdovers from preceding warmer times—dawn redwoods, swamp cypresses, golden larches and maidenhair trees. Dawn redwoods were abundant in wet, low-lying areas, so their foliage makes up the bulk of fossils laid down in swamp and lake sediments. These trees disappeared from North America over 2 million years ago, but living individuals still exist in the wild in a few remote Chinese valleys. These modern trees are so similar to 45-million-year-old fossils that there is a debate among scientists as to whether or not they represent distinct species.

The ancestors of modern conifers were also present, however, having recently descended from the colder highlands—pines, spruces, true firs, sequoias, redcedars, yellow cedars, yews and hemlocks. But the forests had a southern Appalachian touch of broad-leaved trees as well—sycamores, elms, walnuts, beeches, oaks, alders, birches, dogwoods, cherries, magnolias and maples were intermixed with the conifers. Even a small palm *(Uhlia)* was present. And familiar shrubs such as roses, myrtles, grapes, sumacs, elderberries, saskatoons, currants and raspberries formed the rich understorey. So, modern naturalists would feel very much at home walking the game trails of British Columbia 50 million years ago, although they might think they were in the Carolinas rather than northwestern North America.

The lakes themselves were rich in plant, fish and insect life. Waterlilies *(Allenbya)* floated on the surface and swamp willows *(Decodon allenbyensis)* lined the shoreline. Bowfins (similar to the Bowfin of the Mississippi today) and early relatives of the Goldeye were present, but perhaps the most interesting species was *Eosalmo driftwoodensis.* This fascinating fish was a member of the salmon family, intermediate between the salmon-trout group of species and the grayling group. Because there is a full size range of *Eosalma* fossils from fresh water, we can infer that it was not anadromous—that is, this ancestor of our salmon did not migrate to the sea. Familiar fishes missing from this fauna include perches, sticklebacks and sculpins—which all had yet to evolve.

The insects left remarkable fossils. Caddisflies built cases out of small leaves, mayfly larvae fishtailed through the water, and above them, water striders skated about on the surface. A modern species of water strider from British Columbia is virtually identical to these ancient water striders, differing only in the length of the terminal antennal segment of the male. March flies abounded, perhaps an indication of the warmer climate. Other true flies of wet or moist areas, such as crane flies, fungus gnats and wood gnats were also present. Termites, closely related to some living today only in Australia, burrowed in rotting wood, and aphids infested flower heads.

FOSSIL CEDAR FOLIAGE (REDCEDAR OR YELLOW CEDAR) FROM CACHE CREEK

FOSSIL SASKATOON LEAF

EOSALMO DRIFTWOODENSIS FOSSIL FROM THE SMITHERS AREA

an effusion of lava from a multitude of vents in the Chilcotin area from 15 to 2 million years ago. In fact, the immense amount of lava that flowed over the landscape filled in all the low-lying areas and created the flat plateaus of the Cariboo and Chilcotin—all 50,000 square kilometres of them.

Meanwhile, in the Chilcotin, the continental crust was moving westward over a particularly hot spot in the mantle. So a series of shield volcanoes—broad, rounded volcanoes built up by successive outpourings of very fluid lava—formed over the hot spot. As one volcano formed, the crust would carry it to the west and a new volcano would emerge over the hot spot. First the Rainbow Range was formed, followed by the Itchas and the Ilgachuz Range near Anahim Lake and, finally, some very recent cones near Nazko, in the northeastern Chilcotin. The Alert Bay Volcanic Belt on northern Vancouver Island is probably related to deep mantle melting along the northern edge of the Juan de Fuca Plate.

From about 3.5 million years ago to recent times, volcanic eruptions and lava flows have changed the face of the Wells Gray Park area. Visitors to this park can see all sorts of volcanic features, but the most spectacular and popular sight is Helmcken Falls, where the Murtle River cascades over the 142-metre headwall of a canyon carved in a 500,000-year-old lava flow. The origin of the falls can be traced to nearby Pyramid Mountain, which is an 11,000-year-old tuya, or sub-glacial volcano. A short hike to the top of the glass-walled cone produces a magnificent view that reveals how the volcano diverted the Murtle River from its pre-glacial course and sent it over the rim of the canyon. And thus a park is born. The volcanic activity of Wells Gray is related to faults along the eastern edge of Quesnellia, probably caused by shear forces as the terrane is shifted northward.

The most recent activity in the province has occurred in the Stikine Volcanic Belt of the northwest. In about 1750, an eruption dammed the Tseax River in the Nass country, impounding Lava

Lake and destroying villages of the Nisga'a people. The cinder cone of Hoodoo Mountain, near the southern end of the Alaska panhandle, is another volcano believed to be merely biding its time before erupting once more.

Basalt columns form when lava cools quickly. They are a common feature of volcanic landscapes in the Interior and the southern Coast Mountains. These columns are in the Precipice, near Anahim Lake.

TEPHRA: FOLLOW THE WHITE LINES

Some volcanoes erupt quietly and ooze fluid lava; others—those with magma higher in silica—can erupt explosively as Mount St. Helens did on May 18, 1980. Explosive eruptions produce immense amounts of ash—called tephra by geologists—that is blown high into the atmosphere and carried over long distances on the prevailing winds. In the 1980 Mount St. Helens blast, central Washington was smothered in tephra; in British Columbia only a tiny dusting of the greyish-white powder fell in the Okanagan, but more fell in the southern Kootenay region.

Although we can easily see the lava flows produced by volcanoes, most people miss the ash layers that are evidence of the big bangs of the local volcanic world. In British Columbia, there are at least five important layers from such eruptions in the last 10,000 years. The oldest—and still the largest in the region—came from the eruption of Mount Mazama in Oregon about 7300 years ago. Mount Mazama doesn't exist on maps anymore—Crater Lake lies in the caldera produced when the mountain collapsed following the big explosion. Mazama ash covered a broad swath of southern British Columbia (Map 10).

Mount St. Helens erupted 3400 years ago and again 500 years ago, and each time the winds blew significant amounts of ash northeast into southeastern British Columbia. About 2350 years ago, Mount Meager in the southern Coast Range erupted and the ash—known as the Bridge River ash—blew east as far as the Rockies. In the far north, a volcano in the St. Elias Range erupted twice—once 1900 to 1500 years ago and again about 1200 years ago. Ash from the later eruption blew east, and some reached the Atlin district of British Columbia.

These ash layers can be seen as distinct white stripes in road cuts and riverbanks wherever they occur. They are used by geologists to date the sediments that lie immediately below them. Paleoecologists find them especially valuable in dating ancient pollen layers in the peat of bogs and bottom muds of ponds (see the box entitled "Pollen: Dust of the Ages" in Part 2).

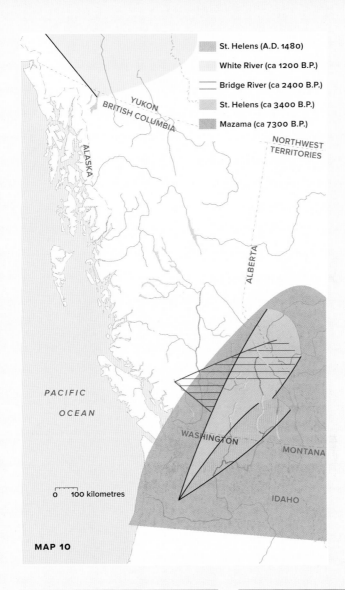

St. Helens (A.D. 1480)

White River (ca 1200 B.P.)

Bridge River (ca 2400 B.P.)

St. Helens (ca 3400 B.P.)

Mazama (ca 7300 B.P.)

YUKON

BRITISH COLUMBIA

NORTHWEST TERRITORIES

ALASKA

ALBERTA

PACIFIC

OCEAN

WASHINGTON

MONTANA

IDAHO

0   100 kilometres

MAP 10

facing page: This line of white volcanic ash in the Similkameen Valley marks the land's surface 7300 years ago, when Mount Mazama erupted and created Crater Lake, Oregon.

MAP 10. VOLCANIC ASH. Layers of volcanic ash in British Columbia. *Adapted from H. Gabrielse and C.J. Yorath, eds., Geology of the Cordilleran Orogen in Canada, Fig. 21.2.*

The Cascade Volcanic Arc and
the Rebirth of the Coast Mountains

The compression that built the original Coast Mountains ended about 45 million years ago, and for the next 40 million years they lay quiescent, gradually eroding down to perhaps a low chain of hills. By 10 million years ago they were so low that there was no rain shadow behind them; the Interior was cloaked in lush vegetation. Then, beginning 5 million years ago, the subduction zone of the Juan de Fuca and Explorer Plates steepened, bringing the subterranean mass of molten rock that had created the Pemberton volcanic belt directly beneath the huge batholithic belt created by the collision of Wrangellia millions of years earlier.

In the south, this change gave birth to the Cascade Volcanic Arc, a series of impressive volcanoes that range from California's Mount Shasta in the south to Mount Garibaldi and Mount Meager in the

facing page: The west face of Mount Garibaldi. The greatest volcanic activity in the Garibaldi region occurred while most of the surrounding land was covered in the great ice sheets of the last glacial age. Mount Garibaldi was born about twenty thousand years ago, as the Ice Age was beginning to wane. The peak rose quickly through the surrounding ice in a series of explosive eruptions and soon became a gently sloping cone of fragmented dacite lava more than 6 cubic kilometres in volume. But to the west, its flank significantly overlapped the glacier, and as the supporting ice melted away, the mountain's entire west face—about 3 cubic kilometres of rock—collapsed into the Cheekye Valley below. In the volcano's last stage of activity, lava flowed gently out of a vent to the north of the previous plug and formed the now slightly higher northern summit.

left: Eve Cone, a postglacial cinder cone on the plateau of Mount Edziza in the Stikine Volcanic Belt. This one of the most recent eruptions of the complex that formed over the past 8 million years, as tension along the edge of the northward-moving Pacific Plate cause rifting inland.

north. In the Garibaldi–Bridge River area, there are actually thirty former volcanic cones extending along a 120-kilometre band.

The heat of the subduction zone not only created volcanoes but also heated the thick crust along the British Columbia coast. As the crust warmed up, it expanded. Geologists calculate that this thermal expansion alone would result in a 2-kilometre uplift of the Coast Mountains. They have, in fact, risen about this much in the last 5 million years—and they continue to rise today.

PLANETARY CYCLES, CLIMATIC CYCLES

Climatic fluctuations are caused by cyclical changes in the earth's orbit and tilt. Three cycles are involved. First, there is a 105,000-year cycle as the earth's orbit changes from a strong ellipse to a more circular shape. Second, there is a 41,000-year cycle in the tilt of the earth's axis—that is, when the earth wobbles in space like a top. Third, there is a 21,000-year cycle in which the timing of the earth's closest approach to the sun moves through the year (right now it is closest to the sun in January).

The primary effect of these three cycles is to alter the contrast in heat between the tropical and polar regions. When the contrast is great, a strong south-north circulation develops and warm air from the tropics brings moisture to the cold north, which falls as snow, and glaciers grow. When the contrast is less, the dominant circulation is east-west and precipitation declines.

These three cycles also influence the contrast between summer and winter. When summers are relatively cool and winters are moderate, there is not enough heat in the summer to melt the snows of winter and glaciers grow. When summers are hot and winters are extremely cold, the heat wins out and mountain snowpacks melt away before they can accumulate into ice.

Where are we in these cycles now? According to some scientists, a long-term cooling trend began about 7000 years ago. There have been shorter-term warm and cool periods within this trend, but the long-term trend was predicted to reach its coolest point 23,000 years from now. But now predictions have to take into account global warming from increased atmospheric carbon dioxide, a trend that could possibly overwhelm the effects of the natural cycles.

Earthquakes: Waiting for the Big One

At 10:00 on Sunday morning, June 23, 1946, the residents of Courtenay were shaken by a tremendous earthquake, registering 7.3 on the Richter scale. The shock was felt as far away as the Lower Mainland, but on the east coast of Vancouver Island rock walls broke, chimneys fell, landslides swept down hills and the seashore subsided. Just three years later, the largest recorded earthquake in Canada—magnitude 8.1—occurred off the northwestern corner of Haida Gwaii.

Earthquakes are an inevitable accompaniment to tectonic plate movement, and British Columbia is no stranger to them. And not all are small, insignificant events—an average of two quakes greater than 6.5 on the Richter scale occur in western Canada each decade. As might be expected, most quakes are directly associated with the Queen Charlotte–Fairweather fault system, where the Pacific Plate slides northward relative to North America (Map 11, overleaf). The southern Strait of Georgia is also seismically active; the earthquakes here are probably related to the subduction of the small Juan de Fuca Plate beneath North America.

Recent surveys by researchers from the Pacific Geoscience Centre in Sidney, British Columbia, indicate that the oceanic crust is no longer sliding smoothly beneath us and that the crust of Vancouver Island is being strained. Mountains are rising and being squeezed closer together. Over the past few years geologists have noticed that the earth's crust in central Vancouver Island is being squeezed at different rates. These changes may very well be the early warning signals of a future earthquake—perhaps the Big One!

> "We have stood atop windswept peaks and watched continents in collision . . . We have travelled to where terranes collide."
> C.J. YORATH, *Where Terranes Collide*

Two hundred million years of plate movement, mountain building relaxation of the crust, volcanic eruptions and earthquakes have created the diverse landscapes of British Columbia. Map 12 (page 61) shows the six major regions formed by all of this geological activity.

The Sculpting of British Columbia: Glaciation

After 200 million years of building and shaping by tectonic movements and water erosion, the stage was set for the sculpting of British Columbia. Great mountain ranges and broad plateaus lay along southeast-northwest lines. The mountains were high but more rounded than today, and they were separated by angular valleys with few lakes and waterfalls.

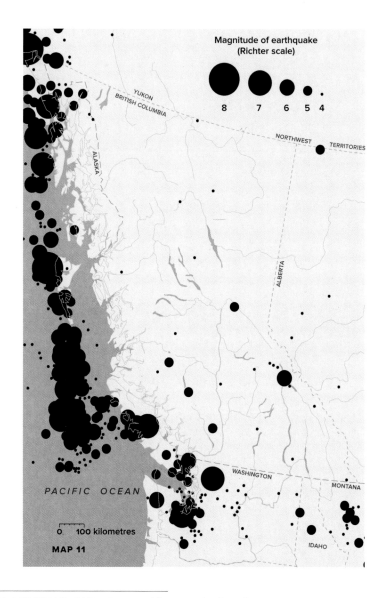

Magnitude of earthquake
(Richter scale)

8   7   6   5 4

MAP 11. HISTORICAL EARTHQUAKES. Historical earthquakes equal to or greater than Richter magnitude 4 in British Columbia and adjacent areas. *Adapted from H. Gabrielse and C.J. Yorath, eds., Geology of the Cordilleran Orogen in Canada, Figs. 21.1 and 21.6.*

Glacial Cycles and Drifting Continents

Then, about 2 million years ago, the age of the great glaciers—the Pleistocene Epoch—began. What caused the ice to come? In several ways, drifting continents had led to a cooling of the earth's climate. First, plate movements had brought the northern continents far enough north to enable glacial ice to grow in colder climatic periods. Second, tropical ocean currents were obstructed by the northward drift of Australia and New Guinea and the closing of the gap between Central and South America at Panama. Finally, barriers to polar currents around Antarctica were eliminated.

But the Pleistocene Epoch was not all cold. It had a series of climatic ups and downs, of moderate climates alternating with distinctly cooler periods. In the last 800,000 years alone there have been twelve of these cycles.

The colder periods were times of extensive glaciation in northern North America. The glaciers sculpted British Columbia by grinding the mountains, by changing the course of rivers and impounding huge lakes and by depositing immense amounts of sediments. Over the last 2 million years glaciers have changed British Columbia and created the face of the province we see today.

At the beginning of each glacial age, either the summer temperature dropped slightly or the precipitation increased or both. In the high country it didn't take much to tip the balance towards the growth of glaciers. Slightly longer, snowier winters produced thicker snowpacks, or slightly cooler summers reduced the melt—either condition meant that snow persisted from one winter to the next.

So the snow began to accumulate—especially in the high basins of the Coast, Columbia and Rocky Mountains, which remain among the snowiest places on earth today. The snow soon built up into small glaciers, which subsequently grew into great valley glaciers flowing onto the surrounding plateaus and lowlands. Eventually, the centre of the province was covered by ice up to 2 kilometres thick. This was the great Cordilleran Ice Sheet, which met the continental Laurentide Ice Sheet east of the Rocky Mountains. Along the border of the Cordilleran Ice Sheet, glaciers in mountains such as the Queen Charlotte Ranges were largely independent of the main ice cap.

Glacial periods ended when the climate became warmer; this climatic change seems to have usually occurred rather suddenly. In the mountains, glaciers retreated back up the valleys, but in the lowlands and on the plateaus, great glaciers were stranded and died in place, melting from the top down.

Between glacial advances, ice-free periods often lasted for a considerable length of time—several lasted 20,000 to 60,000 years. During these periods glaciers were confined to the mountains, as

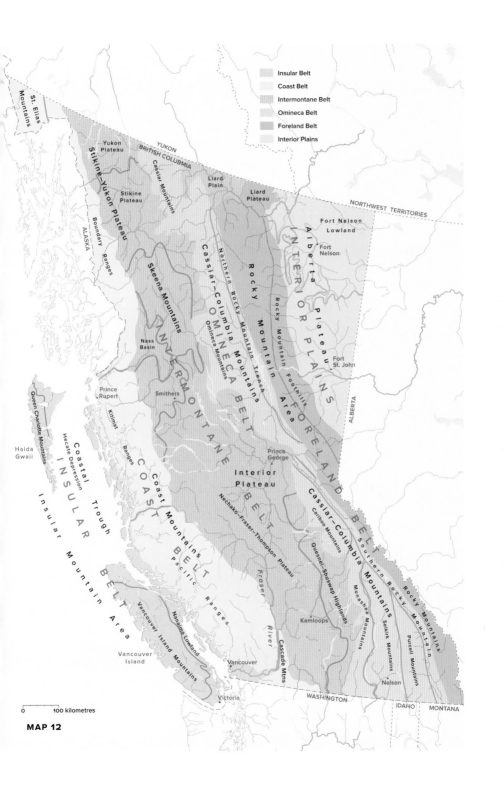

Insular Belt
Coast Belt
Intermontane Belt
Omineca Belt
Foreland Belt
Interior Plains

St. Elias Mountains

Yukon Plateau

YUKON
BRITISH COLUMBIA

Stikine-Yukon Plateau

Stikine Plateau

Cassiar Mountains

Liard Plain

Liard Plateau

NORTHWEST TERRITORIES

ALASKA

Boundary Ranges

Skeena Mountains

Cassiar–Columbia Mountains

Northern Rocky Mountains

Omineca Mountains

OMINECA BELT

Rocky Mountain Trench

Fort Nelson Lowland

Fort Nelson

Rocky Mountain Area

Alberta Plateau

INTERIOR PLAINS

Nass Basin

INTERMONTANE

Prince Rupert

Smithers

Rocky Mountain Foothills

Fort St. John

FORELAND

ALBERTA

Haida Gwaii

Queen Charlotte Mountains

Hecate Depression

Coastal Trough

INSULAR

Kitimat

Kitimat Ranges

BELT

Prince George

Interior Plateau

Cassiar–Columbia Mountains

Cariboo Mountains

Southern Rocky Mountains

Coast Mountains

COAST BELT

Pacific Ranges

Nechako–Fraser–Thompson Plateau

Quesnel–Shuswap Highlands

BELT

Insular Mountain Area

Vancouver Island Mountains

Nanaimo Lowland

Fraser River

Kamloops

Monashee Mountains

Selkirk Mountains

Purcell Mountains

Vancouver Island

Vancouver

Cascade Mtns

Nelson

Victoria

WASHINGTON

IDAHO

MONTANA

0   100 kilometres

MAP 12

they are today, and the valleys and plateaus were recolonized by animals and plants from unglaciated refuges to the south and north (see Part 2). The last major glaciation left the lowlands of British Columbia about 12,000 to 10,000 years ago—we are living in the latest of what could very well be a continuing series of interglacial periods.

Because the more recent glaciers destroyed or reworked the landscape so completely, geologists know little about the earlier glaciations of the Pleistocene. Evidence of one of these glacial advances can be found at the Murtle River bridge in Wells Gray Park—glaciers covering the Clearwater Valley perhaps about 400,000 years ago left behind sediments that were subsequently covered by lava from a volcanic eruption 200,000 years ago.

Geologists do know, however, that there have been two major advances in the last 100,000 years. Because it is older than the limit of radiocarbon dating, the age of the earlier one is not precisely known, but it is definitely older than 59,000 years. Sediments from this glaciation can be seen beneath younger glacial sediments in the Vernon and Okanagan Centre area of the Okanagan and in the lower Fraser Valley. Since its sediments are found beyond the limits of the later glaciation in the central Yukon, we know that the earlier advance was the larger of the two.

facing page: Robson Glacier flows in front of Lynx Mountain in Mount Robson Provincial Park. Like many other glaciers in the Rockies, Robson Glacier has retreated considerably in the past century.

After an interglacial period of at least 30,000 years, the last major glaciation began about 30,000 to 25,000 years ago. The glaciers grew slowly at first, and some areas remained ice free until after 17,000 years ago. The Cordilleran Ice Sheet reached its maximum extent 14,000 years ago but soon began to shrink and, by 10,000 years ago, had vanished, leaving only a few mountain ice fields and glaciers as reminders.

Since this last major glaciation ended, there have been several smaller advances. In fact, the coolest period in the last 10,000 years

was the last few hundred years—the Little Ice Age or, as it is known to local glaciologists, the Cavell Advance, named after Mount Edith Cavell in Jasper National Park. This advance was essentially over by the late 1800s. Since about 1900, the Rockies and the Columbia Mountains have lost about a third of their glacial ice, and most glaciers in the Coast and St. Elias Ranges have retreated.

What Did Glaciation Do to the Landscape?

As the ice built up and flowed out through the valleys, it altered the land in a number of ways (Figure 4). The massive, flowing ice removed the soft bottom sediments of the valleys, and rocky debris held within it carved out the bedrock floors and side walls. In this way, the valleys were deepened, bends were straightened and the valleys were changed from V shaped to U shaped. This valley carving has made it easier for humans to develop transportation routes through mountain ranges in glaciated country. The spectacular fiords along the coast of British Columbia are valleys that were carved below sea level by major valley glaciers. Many of the province's lakes occupy valleys that were similarly deepened where the ice erosion was relatively intense. In fact, Quesnel Lake, at 530 metres deep, is the deepest such lake in the world.

The small side valleys in the mountains were often left "hanging" by the deepening effect of the main valley glacier. These side valleys usually originate in steep-walled mountain bowls called cirques. Cirques are most abundant on the cool north and northeast faces of mountains, where snow accumulates into glaciers that scour the rock at the base and pluck at the walls. The scene is a trademark of British Columbia's mountains—a turquoise mountain tarn lies in the bottom of the cirque's hollow and flows out through an icy stream that plummets from the hanging valley in a postcard waterfall.

Where cirques are carved out on several sides of a peak, they create a steep, spectacular mountain called a horn—Mount Assiniboine

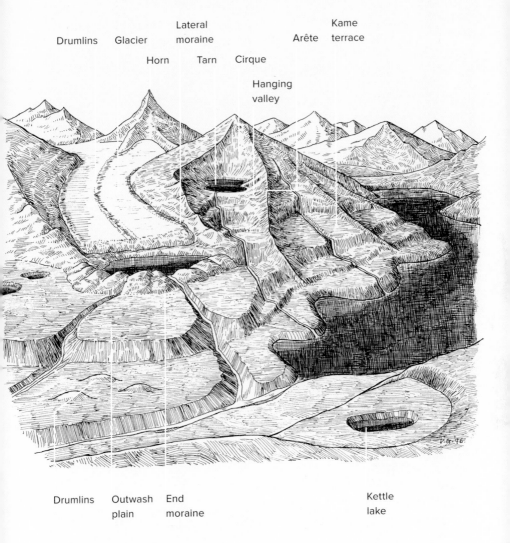

Drumlins Glacier Lateral moraine Horn Tarn Cirque Hanging valley Arête Kame terrace

Drumlins Outwash plain End moraine Kettle lake

FIGURE 4: An imaginary British Columbia landscape reveals the signature of glacial ice on the land.

is a classic example. The knife-edged ridges left between side-by-side cirques are called arêtes.

Often where glaciers override bedrock, the debris within the ice scours scratches and smooth troughs in the underlying rock—glacial striations and grooves that show us the direction of the glacier's flow. Look for these whenever you are hiking in the mountains, but the easiest places for many British Columbians to see them are in the rocks of Lighthouse Park in West Vancouver or below the sea cliffs at Dallas Road in Victoria. A plaque highlights a good example at the Centennial Fountain just west of the Legislative Buildings in Victoria.

The tremendous erosive power of glaciers produced a great deal of sediment during the Pleistocene, and evidence of this erosion is every-where around us. As glaciers move forward over the landscape they act as conveyor belts, carrying within them a great deal of rock and rock debris. This material appears in many forms but is generally referred to as glacial drift. When the glacier stops advancing, the debris builds up at its toe into a ridge called an end moraine. If the toe of a glacier remains relatively stationary for some time, the glacier continues to dump more and more material at the end moraine as it melts each summer. When a glacier begins to retreat, it leaves behind the farthest end moraine, the terminal moraine. Smaller moraines behind the terminal moraine are formed when the glacier pauses occasionally as it retreats—these are termed recessional moraines.

As the glacier melts, rocks within it are also dropped along its sides, creating lateral moraines. When several valley glaciers merge to form one large one, the inner lateral moraines of each merge to form medial moraines, giving the large glacier its typical stripy appearance.

facing page: The rounded grooves in these rocks along Victoria's shoreline were scratched by rock debris frozen within moving glacial ice.

THE PLEISTOCENE SCENE:
BIG ICE, BIG MAMMALS

The Pleistocene glaciations totally reworked the surface of British Columbia and, in doing so, destroyed most traces of the animals that lived here during the warm periods—a number of them tens of thousands of years long—between ice advances. But a few fossils of large mammals have been found that can give us a glimpse into British Columbia between the ice advances. The Pleistocene wasn't all that long ago in geological terms—glaciations began about two million years ago and have come and gone until about 12,000 years ago—but a mammal watcher would notice big differences, even towards the end of the Pleistocene.

There would have been many familiar species, such as Bighorn Sheep, Mountain Goat, Mule Deer, Moose, Black Bear, Grizzly Bear and, along the coast, Northern Sea Lion and Walrus. But they would have been seen with some excitingly different creatures. American Mastodons, twig-eating denizens of marshy areas and open spruce forests, browsed alongside grazers such as Helmeted Muskoxen and Western Bison in open parkland habitats. Imperial Mammoths, standing about 4 metres tall with their immense upward-curving tusks, ruled the tundra world, along with the only slightly smaller Columbian Mammoths. Small horses and Giant Bison also grazed in herds on the grassy tundra. The latter were grandiose animals with a horncore span of over 2 metres. The horncore is the smaller bone inside the horns of animals such as bison—the Giant Bison's horns themselves were not preserved as fossils, so we don't know how large they really were. Hunting the big grazers were huge, lanky Short-faced Bears, which stood 1.5 metres at the shoulder and 3.4 metres on their hind legs. Jefferson's Ground Sloth, a long-haired immigrant from South America that arrived in the Northwest during the last interglacial period, was the size of a modern bear and, like its tropical stay-at-home cousins, stripped leaves from trees for a living. It was first described by

JEFFERSON'S GROUND SLOTH

IMPERIAL MAMMOTH

Thomas Jefferson, the American president, who was also one of the continent's first paleontologists.

The common characteristic of many of these fossil mammals is bigness—mammoth mammoths, large muskoxen, giant bison, huge bears, big sloths and, at least in the ice-free areas of Yukon and Alaska, giant beavers and even giant pikas. They are often referred to as the Pleistocene megafauna. Giant Bison disappeared early, perhaps 30,000 years ago or so, but the Imperial and Columbia Mammoths were grazing on ice-free tundra in southwestern British Columbia about 17,000 years ago. Mastodons roamed with Helmeted Muskoxen and Western Bison only 12,000 years ago. Jefferson's Ground Sloths lasted perhaps until 9000 years ago.

Why did they disappear? There are two schools of thought and, as is often the case when there are two schools of thought, both of them may be right. The first school of thought places most of the blame on climate change. The climate changed dramatically 12,000 years ago.

British Columbia turned from a sea of deep ice to warm earth blanketed in grasslands, woodlands and forests in less than 2000 years. The tundra and cold steppes that originally flanked the great ice sheets shrank dramatically in size. The wide open spaces needed by many of the big mammals such as mammoths became smaller, fragmented spaces. Second, there were humans here then, too—mammoth hunters—and it has been convincingly hypothesized that they could have pushed the mammoths to extinction quickly, especially if the mammoths were already in small, isolated bands. And after the big grazers went, the big predators that preyed on them disappeared too.

HELMETED MUSKOX

SHORT-FACED BEAR

GIANT BISON

At the beginning of a glacial period in the ice ages, the growing glacier ice eroded soft valley-floor sediments, and water from the summer melt flushed out huge volumes of sediments that were redeposited downstream in larger valleys, lakes and fiords. Some glaciers dammed tributary valleys and impounded lakes behind ice dams, in which large quantities of finer sediments—silts and clays—settled out. When the ice cap was at its maximum, most of these sediments were scoured away, but in some areas glaciers overrode them without removing them, and significant older sediments remain.

At the close of a glacial period, more erosion ensued as meltwaters carried off the rocks and sediments loosed by the glaciers. After the ice had gone and forests had returned to the land, the supply of sediment was greatly reduced. Rivers, especially those on steep gradients, quickly cut deeply into the sediments deposited earlier, and present floodplain levels were reached within several thousand years of the glaciers' retreat.

facing page: These elegant white silt bluffs along the Elk River are the bottom sediments of a large glacial lake that occupied this valley at the end of the ice ages.

Where big glaciers languished in the valley bottoms, outwash from the surrounding streams deposited thick beds of gravel and sand against the ice, leaving outwash plains, or flat benches of stratified drift built up against the melting glacier, and kame terraces, or short ridges of stratified drift from the melting glaciers. These terraces are often the only relatively flat land in the narrower valleys of British Columbia and are now prime real estate for either urban development or agriculture.

Where the ice dammed the valley's outlet, meltwater accumulated against the decaying glacier and formed silty lakes on either side of it. The silt slowly settled out onto the lake bottoms. When the glacier finally disappeared and the lakes drained, the silty lake bottoms remained as flat benches on either side of the valley. As their faces erode today, they create the scenic silt cliffs characteristic

MAP 13

0   100 kilometres

YUKON
BRITISH COLUMBIA

NORTHWEST TERRITORIES

ALASKA

Fort
Nelson

Prince
Rupert

Smithers

Fort
St. John

ALBERTA

Haida
Gwaii

Lake Prince
George

Prince
George

Fraser
River

Lake
Kamloops

Kamloops

Lake
Penticton

Vancouver
Island

Vancouver

Nelson

Victoria

WASHINGTON

IDAHO

MONTANA

of the South Thompson, Okanagan and other glacial valleys.

Huge lobes of ice occasionally dammed major rivers, impounding large lakes and forcing meltwaters to empty out through adjacent river systems. At times, the upper Thompson and Nicola Rivers emptied out through the Columbia via the Okanagan system, the headwaters of the Skeena (including Babine Lake) flowed through the Nechako into the Fraser, and some upper Fraser tributaries through the Peace into the Mackenzie (Map 13). The ice-dammed lakes, despite their size, were often ephemeral entities, draining catastrophically when the lake became deep enough to float the ice slightly.

Temporary glacial lakes can still be seen today in the glacier-filled ranges of the St. Elias and northern Coast Mountains. The Alsek River has been dammed repeatedly by advances of the Lowell and Tweedsmuir Glaciers—as late as the mid-1800s, the Lowell backed up the Alsek's waters as far as Haines Junction, Yukon, before the lake drained suddenly, destroying Native villages at the river's mouth in Dry Bay, Alaska. Where the Stikine River slices through the Coast Mountains, the aptly named Flood Glacier dams a tributary of the Flood River and creates Flood Lake. On a smaller scale, the famous Bear River Glacier along the highway to Stewart created an ice-dammed lake (Strohn Lake, not the small meltwater lake at its toe today) at its terminus five times between 1958 and 1962; each time the lake drained, the floods wreaked havoc all the way down the Bear River valley.

The Ice and the Sea

The great ice sheets of the Pleistocene contained so much of the earth's water that the sea level was at least 100 metres lower than

MAP 13. GLACIAL LAKES (*facing page*). Former glacial lakes created by ice dams as the Cordilleran and Continental Ice Sheets melted. The same ice dams temporarily altered the drainage routes of many rivers. These lakes were somewhat ephemeral entities and were not necessarily present at the same time. *Adapted from A.L. Farley.* Atlas of British Columbia, *Map 16.*

Kettle terrain near Riske Creek, west of Williams Lake. Kettles are depressions left when large blocks of ice are stranded among glacial debris. After they melt, they leave hollows in the mantle of drift.

it is today. Much of the coast was covered in ice, of course, but large parts of the continental shelf were exposed during at least the earlier and later periods of the last glaciation. Up to 13,000 years ago the eastern coastal waters of southern Haida Gwaii were dry land and extended more than halfway across what is now Hecate Strait.

But the immense glaciers that covered the British Columbia mainland coast weighed so much that they depressed the land beneath them as much as 250 metres or more. As they melted, the ocean invaded the depressed coastal lowlands vacated by the retreating glaciers. All but the highest points of the Gulf Islands were covered by the sea, as was the entire coastal plain of eastern Vancouver Island. In the Fraser Valley, marine inlets reached from Boundary Bay to at least Pitt Lake and from Bellingham Bay to Agassiz. On the north coast, Pacific fiords occupied the Skeena Valley as far inland as Terrace and the Kitimat-Kitsumkalum Valley almost to Kitsumkalum Lake.

Freed from its icy burden, though, the land began to spring back and the sea level fell correspondingly to its present position relatively quickly—in one or two thousand years. Simultaneously, tremendous amounts of sediment were carried into the valleys from the recently released mountains and plateaus. In the Fraser Valley these two events resulted in the early disappearance of the marine bays, and the Fraser began to build the delta we know today, filling in the shallow bay between Vancouver and Surrey and the island of Point Roberts.

Erosion and Sedimentation Today

"The mighty, muddy Fraser" is a phrase often attached to British Columbia's largest river—and it is an apt one, since each year the Fraser carries about 20 million tonnes of sand, silt and mud from the mountains and plateaus of the Interior and dumps them into the Strait of Georgia. The sediment load of all streams like the Fraser is the result of a myriad of erosional processes—from glaciation to rock-breaking frosts to winter rains to spring floods—that are at work making British Columbia a flatter place.

Studies of the sediment load of rivers on the east slope of the Rockies indicate that these mountains are wearing away at a rate of 6 millimetres every 100 years. From this fact, Ben Gadd, in his *Handbook of the Canadian Rockies*, calculates that Mount Robson will lie at the elevation of Edmonton in exactly 54,766,666 years and 8 months! (But see the box entitled "Erosion, Builder of Mountains" on facing page.)

Although sediments are not piling up as fast as they did immediately following the retreat of the Pleistocene ice sheets, they can still can be a substantial force in creating new landscapes in British Columbia. For example, if it weren't for human interference, the sediments of the Capilano River would someday close off Burrard Inlet and Indian Arm at the Lions Gate Bridge, creating a huge lake

EROSION, BUILDER OF MOUNTAINS

We usually think of erosion as only wearing away mountains, carving them and giving them their rugged faces, but it can be as important as great tectonic collisions in building them as well. It does this by causing local uplift. Like icebergs floating on water, mountains and high plateaus are thick pieces of light crust that float on the denser mantle rock below. As erosion strips away material from the uplands, a mountain can bob back up to 80 per cent of its former height. And if erosion is strongest in the valleys, the mountain can bob up higher than it used to be. In essence, erosion can, then, pull up rocks from far below the earth's surface and thus influence the composition of mountains. In British Columbia, for example, erosion has virtually eliminated the original rocks of the Coast Mountains and, while doing so, has exhumed the range's granitic basement. The granitic rock, originally far below the earth's surface, now soars into the sky, occasionally capped with the remnant rocks of the old mountain range.

Erosion is a function of climate—the more water, the more erosion. In rain-shadow regions, erosion is minor and high plateaus can be created. On the windward side of uplifted areas, erosion is greater and mountains more rugged. In British Columbia, where prevailing winds blow in the same direction in which the Juan de Fuca Plate is being subducted, erosion intensifies the buoyant uplift and exhumation of rock.

in upper Burrard Inlet. But humans have altered the deposition of sediments from the Capilano, just as we have for most of the province's major streams. Dams—such as the Cleveland Dam on the Capilano—allow sediments to fall out of rivers into the reservoirs, reducing sedimentation at their mouths. Sedimentation in reservoirs is a real problem since it will eventually fill the reservoir and render the dam useless.

The Capilano's estuary has also been greatly altered. The river has been diked and its mouth moved to the west of the Lions Gate Bridge footing. When rivers are diked, sediments are no longer spread out over estuaries and floodplains in the annual spring floods—they instead go right out to sea and are dumped at the edge of their deltas. On the floodplain and delta of the Fraser River, 530 kilometres of dikes have been constructed and 70,000 hectares of marshland and floodplain have been converted into agricultural and urban land.

ROCKS AND DIRT:
THE FOUNDATION OF LIFE

Soil is a product of the climate above it, the plants within it and the rock below it. The bedrock or sediments below are the raw materials that the vegetation and climate work on to produce, over time, productive soil. As different types of rocks weather, they produce different types of soils and, ultimately, different types of vegetation.

Although most soils in British Columbia are derived from glacial drift and sediments, observant naturalists can often see the effect of bedrock on soils and vegetation in the mountains. Soils over volcanic rocks are usually deep and rich in nutrients. Shale weathers into muds and silts, which make up the valuable soils of river floodplains and deltas. Granitic rocks, which make up most of the Coast Mountains, weather into poor, acidic soil. Serpentine rocks, which are impoverished in essential elements such as calcium, potassium and phosphorus, produce soils that only a few plants—some of which are unique to these sites—can tolerate.

Limestone, which is primarily made up of calcium carbonate, weathers by

dissolving almost completely and so tends to accumulate a soil mantle very slowly. But because the carbonate neutralizes acids, distinctive plant communities grow on limestone bedrock. The next time you are hiking in limestone-rich mountains such as the Rockies, take a look at the boundary between limestone and shale formations. The shale produces acidic soil covered by dwarf shrubs such as crowberry and mountain-heathers, whereas the limestone tundra is dominated by grasses and sedges.

The colder and wetter the climate, the less influence that underlying geology has on the soils above, since the water tends to leach all the nutrients away and leave an acidic, nutrient-poor soil.

The limestone rocks of the Cache Creek terrane, seen here at Doc English Bluff southwest of Williams Lake, weather to calcium-rich soils that are host to a number of unusual plants.

To keep channels open for large boats, rivers must be dredged—every year, 2 to 4 million tonnes of sediments are removed from the mouth of the Fraser. The removal of sand from the Fraser and engineering work around the Point Grey cliffs have affected the beaches of Greater Vancouver in a very direct way, resulting in a shortage of sand feeding onto Spanish Banks.

Much of British Columbia's lowlands are mantled with sediment laid down as glaciers melted. Here, pebbles and cobbles are bedded in a terrace beside the Fraser River near Big Bar.

The Living Land

Although the land around us appears stable and fixed, it is obvious that it has changed remarkably over the eons and continues to change today (Table 1, pages 14–15). In a sense the land is living; continents move, ocean floors are swallowed up, and mountain ranges rise and fall. It is difficult to perceive the change over a human lifetime, but the evidence of the moving land is in the rocks we stand on. This grand geological history is the foundation for the natural world of British Columbia, and our enjoyment and appreciation of nature are increased immeasurably by understanding it.

PART TWO
The Legacy of the Ice Age

ODAY THE GREAT ice sheets of the Pleistocene are gone, but their legacy remains. Their advances and retreats not only changed the physical landscape utterly but also shaped the plant and animal communities that followed them.

Post-Glacial Immigration

Most of the plants and animals of British Columbia are descendants of immigrants that colonized British Columbia after the retreat of the Pleistocene ice sheets only 10,000 years ago. How they populated the landscape left bare by the shrinking ice is a complex, fascinating story. Although all of the province's species have particular ecological needs that determine where they *can* live, where they *do* live relates just as much to where their forebears lived during the Pleistocene, the immigration routes of their forebears following the glaciers' retreat and their ability to disperse. For some species, colonization continues today.

The Role of Glacial Refuges

During the glacial advances of the Pleistocene, there were four primary ice-free areas where species from British Columbia survived: (1) the forests, grasslands and shoreline habitats of coastal California, Oregon and southern Washington; (2) the tundra, forests and grasslands of the interior southwestern United States; (3) the forests and grasslands of the southeastern United States; and (4) the

tundra and cold steppes of central and northern Yukon and Alaska
(Map 14, overleaf). As they followed the retreat of the glaciers, some
species spread across British Columbia, but many were thwarted
by barriers of mountain ranges, ice caps and wide rivers. The more
restricted distribution of these species today can be directly related
to their home during the ice ages. For example, some songbirds such
as the Cape May Warbler emigrated from the southeastern United
States but never managed to cross the Rocky Mountains—in British
Columbia, species such as this are found only in the Dawson Creek
and Fort Nelson areas.

Species that were widespread in the forests of northern North
America were split apart by the Pleistocene ice sheets. Forced south
by the cooling climate and advancing ice sheets, eastern and western populations were separated by inhospitable, treeless plains, and
this separation had profound effects on their evolution. Not only
were major eastern and western groups separated from each other,
but many populations in the west were fragmented into still smaller
groups by mountain ranges and desert basins.

In a small, isolated population, random genetic changes that
begin in an individual can become firmly established throughout
the population over relatively few generations, even without the
pressures of adaptation and natural selection. This means that
chance differences in characteristics such as appearance can evolve
rapidly. And because the isolated populations are living in different

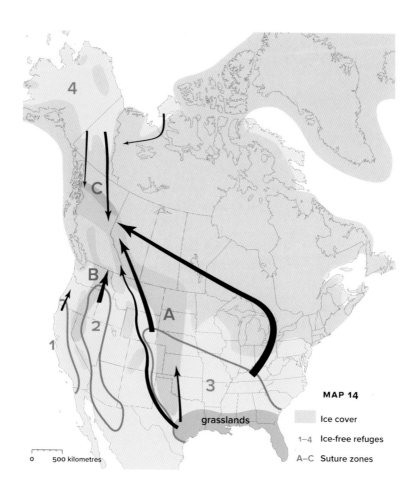

MAP 14. ICE-FREE REFUGES. Ice-free refuges for British Columbian animal and plant species during maximum glacial periods of the Pleistocene. Recolonization routes following the retreat of the ice sheets are shown by arrows. Also shown are biological suture zones entering British Columbia: (A) Northeast-Southwest; (B) Coastal-Interior; (C) Beringia-Cordillera. Each of these zones represents the coming together of many formerly separated populations of animals and plants. *Adapted from H. Danks, ed.,* Canada and Its Insect Fauna *(Ottawa: Entomological Society of Canada, 1979), Fig. 3.45.*

TABLE 2. EAST-WEST PAIRS OF PLANTS AND
ANIMALS OF THE NORTHERN FOREST

Note: Subspecies names are in parentheses

East	West
White Spruce	Engelmann Spruce
Jack Pine	Lodgepole Pine
Balsam Poplar	Black Cottonwood
Yellow-bellied Sapsucker	Red-naped/Red-breasted sapsuckers
Northern (Yellow-shafted) Flicker	Northern (Red-shafted) Flicker
Yellow-rumped (Myrtle) Warbler	Yellow-rumped (Audubon's) Warbler
Rose-breasted Grosbeak	Black-headed Grosbeak
Dark-eyed (Slate-colored) Junco	Dark-eyed (Oregon) Junco
Baltimore Oriole	Bullock's Oriole

places, their appearance, behaviour and ecological requirements often change through *adaptation* to new habitats or climates as well.

For example, the ranges of species of the northern forests, such as the White Spruce, were split in two during glacial periods. Forced south by the advancing ice sheets, eastern and western populations were separated by inhospitable, treeless plains. Over the centuries, they evolved different appearances and, to some extent, different ecological strategies. Examples of east-west pairs abound in British Columbia (Table 2). Jason Weir and Dolph Schluter of the University of British Columbia studied a number of species pairs of boreal forest birds, and, using a standard genetic mutation "clock," they showed that all of these northern pairs diverged during the Pleistocene ice ages.

CAPE MAY WARBLER

Windblown Whitebark Pine hangs on atop a mountain pass in the Charlotte Alplands south of Anahim Lake. British Columbia's mountain chains act as barriers separating coast and interior forms of lowland species.

With few exceptions (the Jack and Lodgepole Pine pair being one), the "eastern" member of these pairs is today the northern member and the "western" member is really the southern, mountain one. The eastern Yellow-shafted Flicker, for example, is found right across northern British Columbia and, in Alaska, extends much farther west than the western Red-shafted Flicker. This fascinating, consistent pattern is the result of the timing of deglaciation in different parts of the Canadian west. As the Pleistocene ice sheets melted, the first corridor to open up the north and connect ice-free southern North America with ice-free Yukon and Alaska was probably along the thin edge of the continental sheet on the east slope of the Rocky Mountains. This corridor was invaded early by White Spruce and its eastern companions. As the

ice continued to melt, these species moved through the low passes in the northern Rockies and colonized northern British Columbia. Their western siblings, however, were isolated from the north by huge ice domes that persisted in central British Columbia. These populations remain south of Prince George even today—an ecological case of first come, first served.

West of the Rocky Mountains, many forest animals and plants were separated into coastal and Interior populations by harsh, dry conditions in the Interior northwest during the Pleistocene glaciations; today these groups come into contact along the crest of the Cascade Range and Coast Mountains. Examples include the Interior and coastal subspecies of Douglas-fir, Red and Douglas Squirrels, Red-naped and Red-breasted Sapsuckers, and Cordilleran and Pacific-slope Flycatchers. In some species, these groups exist but are not obvious to the average observer. Genetic analysis of Black Bear,

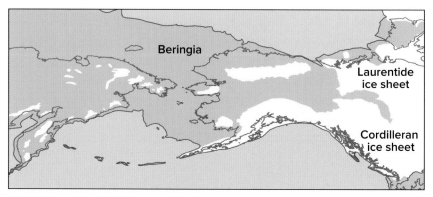

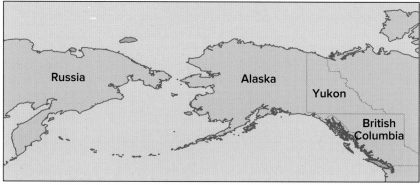

MAP 15. BERINGIA. *Top*: Ice-free land (green) at the height of the Pleistocene glaciations. Beringia is the land around the present-day Bering Strait. *Bottom*: Present-day geography.

American Marten and Montane Shrew all reveal coastal and Interior populations that have diverged genetically but not, to our untrained eyes, in appearance.

Some coastal-Interior species pairs, such as the sapsuckers mentioned above and the Hermit and Townsend's Warblers (see the box entitled "Genetic Genocide" on page 106), also have an eastern sibling. In these cases, the two western species diverged from their eastern sibling first—perhaps a million years ago according to Weir and Schluter's DNA clock—and subsequently diverged from each other later in the Pleistocene.

To the north, another group of animals and plants flourished in the cool but ice-free lands of the Yukon, Alaska and eastern Asia. During the height of glaciation, so much of the world's water was tied up in ice that sea level was about 100 metres lower than it is now. The unglaciated portion of Alaska and the Yukon was broadly connected by the Bering land bridge to eastern Siberia. Totally isolated from the rest of North America by immense glaciers, this area (termed Beringia by geographers) was biologically more Asian than North American during the Pleistocene (Map 15).

MOUNTAIN HAREBELL

The plants and animals that lived in Beringia were citizens of the tundra and cold steppe, and when the intervening glaciers melted, many species dispersed down the alpine spines of the Rockies and the Coast Mountains. The White Marsh Marigold, Mountain Monkshood, Partridgefoot, Mountain Harebell and Golden Saxifrage probably all populated British Columbia's mountains from the north—although some might have survived in high mountain refuges within British Columbia. Grizzly bears swept out of Beringia to populate all of western North America.

MOUNTAIN MONKSHOOD

Just as some species were divided east and west by glaciation, others were split into populations north and south of the ice sheets. In British Columbia, some Beringian species are still confined to the far northwest. Examples of Beringian-southern pairs include Harlan's and Red-tailed Hawks, Collared and American Pikas, Arctic and Columbian Ground Squirrels, and Thinhorn

PARTRIDGEFOOT

(Dall's) and Bighorn Sheep. Even spawning populations of salmon were separated; Chinook Salmon diverged into a stream-type fish—which enters fresh water in the spring—north of the glaciers and an ocean-type fish—which enters fresh water in the fall—to the south. Dolly Varden split into two races as well, although in this case, the northern type is confined to streams that enter the Bering and Beaufort Seas, far to the north of British Columbia.

A number of species pairs now meet and hybridize along an east-west zone stretching through Prince George. This is a meeting of two biotas, groups of different types of plants and animals, that were once separated but are now in contact again. These zones of hybridization are called suture zones, and they provide vital biological field laboratories for studying the great changes that have occurred over the last 100,000 years.

Three of the suture zones that snake across North America make their way into British Columbia (Map 14, page 84). The longest is the zone mentioned above, where northern and eastern populations meet their western counterparts. This zone follows the hundredth meridian north along the boundary of the humid eastern woodlands and the dry western plains and then swings west to the crest of the Rockies, continuing north before swinging west through central British Columbia. Another zone follows the crest of the Cascade and Coast Mountains, and a third zone marks where plants and animals of the far northwest meet their siblings of the southern mountains.

Some species pairs hybridize extensively in suture zones; others hybridize much less readily or do not even come into contact. Whether or not these pairs are new, separate species is a matter of some debate and depends on the individual circumstances as well as the various specialists' concept of what a species is. For example, Yellow-shafted and Red-shafted Flickers hybridize so widely that ornithologists generally agree that they are only well-marked, geographic subspecies.

White and Engelmann Spruce hybridize widely, but many botanists still treat them as separate species, calling most spruce in central British Columbia merely hybrid spruce. Myrtle and Audubon's (collectively Yellow-rumped) Warblers also meet near Prince George, but there is only a very narrow zone where hybrids occur. Recent genetic research indicates that these two are best treated as separate species as well. Baltimore Orioles and Rose-breasted Grosbeaks range into the Peace River district, but their western counterparts, Bullock's Orioles and Black-headed Grosbeaks, do not extend that far north. Thus the two groups fail to meet there, though there has been a single report of hybrid grosbeaks in the Northern Rocky Mountain Trench near McBride.

BALTIMORE ORIOLE

BULLOCK'S ORIOLE

Although most of British Columbia was completely overrun with ice during the Pleistocene, there is considerable evidence that a number of small areas in British Columbia remained ice free during the last glacial advance. The peaks of the highest mountains poked above the glacial sheets below, but these were inhospitable places unlikely to support much life.

Along the outer coast, however, especially on the outer coasts of Vancouver Island and Haida Gwaii, there were more hospitable refuges. Where the Cordilleran glaciers tapered off into ice shelves, even modestly elevated ridges were exposed above the ice surface. And between the tongues of large glaciers, steep west-facing seaside slopes often escaped the ice and supported communities of plant and animal life. Because sea level was considerably lower during maximum glaciation, these seaside refuges would have been significantly larger than the corresponding slopes are today. They would have been even more extensive farther out on the present

MIXED-UP FLICKERS

A glance at an old field guide for birds will reveal one of the classic east-west species pairs: the Yellow-shafted Flicker of the east and the Red-shafted Flicker of the west. The plumages of these two ground-feeding woodpeckers are very different: eastern birds have bright yellow underwings and tail, a brown face and grey crown and a red crescent on the back of the head, and the males have black whiskers; western birds have salmon-red underwings and tail, a grey face and brown crown and no crescent on the back of the head, and the males have red whiskers. Because they look so different, one might expect that the eastern and western birds would shun each other completely. But perhaps because feeding behaviour and mating displays are essentially identical between eastern and western populations, flickers from one population do not hesitate to mate with members of the other population. Hybrids are very common across central British Columbia, where the two populations meet, and birders in southern British Columbia often encounter them in winter. These confusing birds have yellow-orange wings (or both red and yellow feathers), black and red whiskers and other mixed-up characteristics in all imaginable combinations. Do not expect these garbled characters to spread out of the central plateaus; the large pool of pure flickers in northern and southern British Columbia will keep the hybrid zone stable for millennia.

Red-shafted Flicker, the southwestern representative of the Northern Flicker.

Red-breasted (*above*) and Red-naped (*below*) Sapsuckers meet along the crest of the Cascade and southern Coast Mountains. A good place to look for hybrids is Allison Pass, along Highway 3 in Manning Provincial Park.

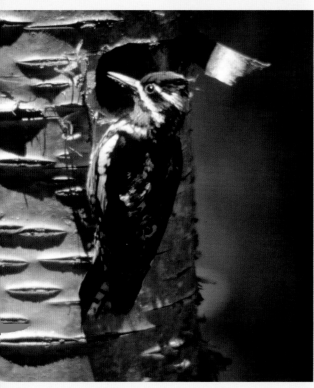

continental shelf, where the ice was thin and the land had not been depressed as much by the ice.

Traditionally, scientists have looked to endemic species as evidence for the existence of these refuges. Endemics are species that are confined to small regions and are found nowhere else in the world. There is a group of wildflower species, for example, that is found only on the ridges of Moresby Island on Haida Gwaii and on the rugged spine of the Brooks Peninsula on northwestern Vancouver Island. Flowers such as Taylor's Saxifrage and Newcombe's Butterweed appear to be remnants of a much more widespread coastal floor eliminated elsewhere by glaciation. These species, which obviously have not dispersed very far in the last 10,000 years, are biological markers that record the areas that survived glaciation. Another similar pattern occurs in some alpine plants that are restricted to the mountains of Vancouver Island and the Olympic Peninsula of Washington State.

Other endemic species of plants—and beetles as well—are restricted to gravelly or rocky seaside habitats along the east coast of Haida Gwaii and the adjacent mainland coast. This pattern points to refuges at sea level, and there is direct evidence of alpine communities in plant remains held in sea cliff exposures on Graham Island that have been dated at 16,000 to 13,000 years ago. Half of Hecate

Strait was once dry land with tundralike meadows, but these in fact might have been post-glacial.

These refuges must have been ephemeral, however, since these shores of the island show evidence that they were covered with ice at some point during the last ice advance. But at any one time, there was probably a chain of ice-free refuges along the outer west coast of North America. Although they came and went, these refuges acted as steppingstones for animals and plant species moving up and down the coast.

In the last few years, new genetic techniques have made it much easier to study the genetic legacy of animals and plants, and the stories coming out of British Columbia do carry signatures of Pleistocene refugia, confirming those told through the distribution of species. Lodgepole Pines, for example, carry the genetic signature of five refugia: Beringia in the north, the outer north coast, the coast south of the ice sheets, the Interior west of the Rockies and the Rockies themselves. Pines growing in northern British Columbia and southern Yukon carry the signatures of both the southern Interior and Beringia. There, the southern trees seem to have moved into the dry valleys and low plateaus, and the northern trees are now found primarily in the subalpine zone.

A word of caution: because the Pleistocene was an extremely dynamic time for animals and plants, with numerous glacial advances interspersed with retreats and relatively long periods of amenable climate, it can be tricky to tease out the historical causes of today's patterns. One example would be the story of the Grizzly Bear. If one were to look at the genetics of today's Grizzly Bears, one would see a southern group with its unique mitochondrial DNA signature, a north coastal group with its special signature, a Yukon group with a Beringian signature and an Alaskan group with an Asian signature. This pattern would suggest that one group originated south of the ice sheets, another originated in coastal refugia, another was isolated in Beringia, and the Alaskan group emigrated recently from Asia. But there is no fossil evidence of bears inhabiting southern North America during the Pleistocene. What can it mean? We would still be scratching our heads, except that recent analysis of the DNA of Pleistocene bears frozen in the permafrost of Alaska and the central Yukon show *all* of these DNA signatures present in that area! It seems that, during the Pleistocene, the population inhabiting Beringia was far more genetically diverse than today's populations and that the small populations emigrating out of Beringia each carried only part of that diversity, resulting in today's differences.

facing page: There is no fossil evidence of Grizzly Bears south of the Pleistocene ice sheets in North America—they moved south into the plains and Rocky Mountains only after the ice had left.

Evolution of Rivers and the Fish Invasion

How did freshwater fish repopulate drainages such as the Fraser and the Skeena, which were totally glaciated only 14,000 years ago? Anadromous fish such as salmon, Steelhead and White Sturgeon had no problems—they moved in from the ocean, swimming up the young, silt-laden rivers to find new pools and new spawning gravels. Fish entirely restricted to fresh water, however, had to emigrate from river systems that had remained at least partly ice free during the Pleistocene glaciations—systems such as the Chehalis, the Columbia, the Missouri-Mississippi and the Yukon.

Fish can't normally cross passes between watersheds—but they had to move across several divides to populate the Skeena from the Columbia. How did they do it? The answer lies in the complex interaction of evolving river systems and the retreats and advances of glaciers at the close of the Pleistocene. In simple terms, the rivers acted as canals—as local glaciers advanced, they closed and flooded the rivers' locks, sending the water from one drainage into the neighbouring one over now-elevated divides.

The only solid piece of scientific truth about which I feel totally confident is that we are profoundly ignorant about nature.
LEWIS THOMAS, *The Lives of a Cell*

While glaciers covered British Columbia and Puget Sound, a unique fish fauna—closely related to that in the Columbia— developed in the isolated Chehalis River at the base of the Olympic Peninsula. As the massive ice lobe melted back up Puget Sound, large lakes formed along its margin. The Chehalis fish entered these lakes and followed their meltwater streams north into the Fraser Lowlands. But the Fraser Canyon was blocked by ice until about 11,500 years ago, so the Chehalis fish were able to establish themselves there before any fish came down the Fraser from the north.

Two fish that represent this interesting fauna in British Columbia are the Salish Sucker and the Nooksack Dace, both now restricted to a handful of small streams in the lower Fraser Valley and both gravely threatened by urban and agricultural development. Bull Trout also survived in the Chehalis system, and the descendants of that population managed to colonize the Squamish River drainage with their close cousins, the anadromous Dolly Varden. All other Bull Trout populations are confined to Interior drainages.

Large lakes were also dammed by glaciers retreating from the valleys of the southern Interior (Map 13, page 72). For a time, the upper Thompson and Nicola Rivers drained through the big lakes in the Thompson, Shuswap and Okanagan systems and into the Columbia. So the fish in the Columbia system easily entered the waters of the Fraser system as soon as the ice melted.

A little later, an ice advance dammed some upper tributaries of the Fraser and forced them to drain north over the pass at Summit Lake and into the Peace system. The connection was brief, but it did allow some Columbia River fish to enter the Peace River.

Around the same time, the Skeena Valley was blocked by glaciers flowing out of the Coast Mountains, and the upper Skeena, including the Babine River, drained east into the Nechako and upper Fraser system. In this way, some species of Columbia fish colonized the Skeena.

WHITE SUCKER

The continental ice sheet left the eastern foothills of the Rockies relatively early, and as it did, a complex of large lakes covered much of the Peace River country. At different times, these lakes drained either south into the Missouri-Mississippi system or north into the Mackenzie system, allowing Missouri River fish to colonize most of northeastern British Columbia.

Before about 10,000 years ago, a lake briefly covered the Peace River Canyon, which before had been a serious barrier to fish movement. This lake allowed some Missouri fish to enter the upper Peace and later, via the Summit Lake connection near Prince George, allowed two Great Plains species—the White Sucker and Brassy Minnow—to colonize the Fraser.

The most fascinating story about fish in British Columbia is the rapid evolution of new forms following deglaciation. Until recently, students were taught that evolution occurs gradually over the grand scale of geologic time and that the origin of a new species spans several million years. Recent discoveries have swept those ideas aside, however. Two conditions are needed for rapid evolution—geographical isolation and ecological opportunities—and for freshwater fish, post-glacial British Columbia was an evolutionary paradise.

Fish entered various river systems through huge but ephemeral lakes and were subsequently isolated from their parent populations. The fish communities in the new lakes and rivers did not have to cope with the diversity of species that their ancestors had coped with south of the ice, so there was a variety of new ecological opportunities available.

New forms appeared quite quickly—a tiny deepwater sculpin in Cultus Lake, landlocked smelts in Pitt and Harrison Lakes, "giant" Pygmy Whitefish in McLeese and Tyhee Lakes, an unusual Largescale Sucker in the upper Kettle River, a lamprey that never leaves Cowichan Lake and a species pair of whitefish that existed in Dragon Lake before the fish were poisoned so that trout could be introduced. And the list could be a lot longer. Our knowledge of fish evolution and taxonomy in British Columbia is limited, and more detailed studies will undoubtedly uncover many more such cases.

One of the best examples of fast-tracked evolution in the world is the story of the threespine stickleback and its colonization of the west coast of British Columbia. The threespine stickleback is a spiny little fish that is common along the coasts of the North Pacific and North Atlantic Oceans. It feeds on plankton in shallow marine waters but also invades creeks. Along the glaciated coast of British Columbia and Alaska, many lowland lakes contain stickleback populations.

In six lakes on the Gulf Islands and Vancouver Island, however, a curious event has occurred. In each of these lakes (Spectacle, Paxton, Priest and Emily Lakes on Texada Island; Hadley Lake on Lasqueti Island; and Enos Lake on Vancouver Island), there are two species of threespine stickleback. One lives near the edge of the lake, feeding on invertebrates in the weeds and mud of the lake bottom; the other is a fish of open water, feeding on the plankton in the water column. But although each pair shares its lake in a similar ecological way, there are differences between the corresponding species in different lakes, implying that each lake's pair has evolved separately.

What does glaciation have to do with this phenomenon? First, since all these lakes were covered with ice during the last glaciation,

OPEN-WATER STICKLEBACK SPECIES

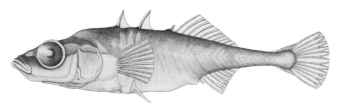

BOTTOM-FEEDING STICKLEBACK SPECIES

these species must have evolved in the last 12,000 years. The simplest explanation of this phenomenon involves the intriguing events that followed the melting of the glaciers along the coast.

The immense ice sheets that covered the Strait of Georgia and Vancouver Island compressed the land beneath them. As the ice melted, the salt water invaded the low-lying land exposed by the retreating glaciers. All but the highest points of the Gulf Islands were covered by the sea, as was the entire coastal plain of eastern Vancouver Island. The sticklebacks came along with the invading sea and populated the new inlets and basins. Then, freed from its icy burden, the land sprang back, and the sea level quickly fell correspondingly to its present position.

As the sea level fell, populations of sticklebacks were left behind in new lakes and were isolated from their sea-going relatives by waterfalls and rapids on the new coastal creeks. In big, steep-walled fiordlike lakes, the sticklebacks remained general plankton feeders like their ancestors. And they remained similar in appearance too— slim, with small mouths and long gill rakers. But in lakes with broad, shallow, warm edges rich in food, such as the six lakes named above, they began to exploit the invertebrate life on the muddy, weedy bottom. This successful shift in feeding strategy selected for individuals with a noticeably different form from that of other sticklebacks, and populations of stocky, wide-mouthed fish with short gill rakers soon evolved.

Then, 1000 to 2000 years after the sea level had fallen, it rose again, although only to about 50 metres above its present level. This rise was enough, however, to carry marine sticklebacks over some of the barriers separating them from the lowland lakes. Some of them reached the lakes colonized many generations before, but when they got to lakes with bottom-feeding sticklebacks present, the immigrant fish found that they couldn't compete with the residents. They could

Mountain Goats lived in British Columbia 90,000 years ago, as fossil bones found in interglacial deposits attest. After the last glacial period, these hardy animals were probably among the first to recolonize the ice-free mountain slopes.

only coexist with residents if the lake was relatively large and deep—that is, if it had a well-developed open-water zone with enough plankton for the new arrivals to live on.

As the sea level fell once more, the lakes were again isolated from marine sticklebacks. Even though the two populations in the lake were closely related and could produce fertile hybrid offspring, they rarely interbred—and the differences between them were maintained generation after generation. They had essentially become new, separate species—in less than 2000 years.

It turns out that this phenomenon occurred in only a handful of lakes in the Strait of Georgia area. These lakes were all low enough to be colonized by two consecutive immigrant invasions, high enough to become isolated from the ocean by waterfalls or other barriers and large enough to have a substantial open-water zone. They also had broad, sloping edges that were shallow enough to support a rich littoral, or shoreline, zone.

Who needs to go to the tropics to look for exotic stories of evolution in action when we have such superb examples right here in British Columbia? These little sticklebacks are biological treasures—and we should treat them with the care due to unique creatures. Tragically, we have already lost one pair of species—predatory catfish were introduced into Hadley Lake and have apparently caused the extinction of the sticklebacks there.

Island Hopping

The coast of British Columbia contains one of the great archipelagos of the world: thousands of islands, ranging in size from the great Vancouver Island to small rocky islets that barely make it onto the marine charts. Most of these islands were glaciated up to 12,000 years ago, but now they abound with life. A closer look reveals that their plant and animal communities are not the same as those on the mainland.

GENETIC GENOCIDE

Recent genetic research by Sievert Rohwer of the University of Washington has uncovered a fascinating story of recolonization and genetic genocide in a pair of wood warblers. The Townsend's Warbler is one of the commonest warblers in the coniferous forests of British Columbia's coast and Interior mountains. The closest relative of the Townsend's Warbler is the Hermit Warbler, which lives to the south of British Columbia—in the Coast, Cascade and Sierra Mountains of California, Oregon and southern Washington. The ancestor of these two species was apparently split from the wide-ranging, boreal stock of the eastern Black-throated Green Warbler during one of the glacial advances of the ice ages.

TOWNSEND'S WARBLER

Today, these species meet and hybridize in two areas of Washington State: in the Olympic Mountains and in the Cascade Mountains south of Mount Rainier. Closer examination of birds around the contact zone indicated that genes of the Townsend's Warbler infiltrate the populations of Hermit Warbler to the south, but the reverse is not true; there is no evidence of Hermit Warbler genes in Townsend's Warblers north of the contact zone. Sievert Rohwer decided to investigate this hybrid zone by looking at the mitochondrial DNA of the warblers. Mitochondrial DNA (mtDNA) isn't involved directly in sexual reproduction but is passed on from mother to offspring in minute intracellular organelles within the egg. It is therefore a good marker of maternal lineages.

Rohwer's studies revealed a remarkable pattern. The Townsend's Warblers of the Rockies had "pure" Townsend's mtDNA, and the Hermit Warblers of the Sierras had "pure" Hermit mtDNA as predicted, but the Townsend's of the British Columbia coast had a good portion of Hermit Warbler DNA in their mitochondria. Even more remarkable, the Townsend's of Haida Gwaii and Prince of Wales Island (of the Alaska panhandle) had 100 per cent Hermit Warbler mtDNA!

What does this tell us? It tells us that at one time, Hermit Warblers inhabited the entire west coast of British Columbia—and that they have subsequently been genetically overrun by Townsend's Warblers.

How did this happen? It seems that the ancestors of Townsend's and Hermit Warblers were separated from each other during glaciation, just as they were both separated from their boreal counterparts. They were isolated by the unfriendly desert habitat of the Great Basin, and the Rocky Mountain group evolved into the

Townsend's Warbler and the coastal group evolved in the Hermit Warbler.

As the glaciers retreated, Townsend's Warblers moved up the Rockies into the Interior of British Columbia but were blocked from the coast by the still-extensive ice dome of the central Interior. Hermit Warblers colonized the newly created temperate coniferous forests of the coast, reaching well into coastal Alaska. But about 5000 years ago, when the ice finally retreated to the highest mountain peaks and coniferous forests moved through the large valleys, Townsend's Warblers descended on the coast and began hybridizing with Hermit Warblers. But how did they take over?

Sievert Rowher has a theory, a story that explains that. Part of it is simple timing, and part is simple aggression. First of all, we must remember that the mtDNA is a record of the maternal lineage, so most of the matriarchs of the coastal Townsend's Warbler line were Hermit Warbler females. Each spring, male Townsend's Warblers arrive on the breeding grounds a few days earlier than the females so that they can establish territories before the females appear. But Hermit Warblers migrate somewhat earlier than Townsend's Warblers, so when the male Townsend's arrived on the coast, female Hermit Warblers were already there, ready to mate with attractive males. Even though they looked a bit odd and sang somewhat differently, the male Townsend's must have been quite acceptable to the female Hermit Warblers. So the male Townsend's took up with the female Hermits. But what about the male Hermit Warblers—did they just stand by? Well, in a manner of speaking, yes. It turns out that by various measures, male Townsend's

Warblers are four times as aggressive in establishing territories and securing mates as male Hermit Warblers! Sievert Rohwer speculates that this difference is a product of the very limited breeding habitat that Townsend's had in the Rocky Mountains at the height of the ice ages, in comparison with the vast coastal forests available to Hermit Warblers. Breeding habitat was never limiting for Hermit Warblers, so the males never had to really fight for their territories.

And so each spring the story repeats itself—male Hermit Warblers continue to lose ground to their invaders, and the Townsend's Warbler hegemony moves ever southward. The genetic ghosts of the Hermit Warblers, though, remain and will remain for many thousands of years.

VANCOUVER ISLAND MARMOT

facing page: Many mammal species familiar to mainland naturalists, such as the Northwestern Chipmunk, have never made the short crossing to Vancouver Island.

Consider Vancouver Island. It was likely revegetated quickly with wind-borne seeds, and most birds made the flight over from the mainland in a matter of geological minutes (although one species, the Black-capped Chickadee, familiar to most back-yard birdwatchers on the mainland, is absent to this day). Mammals, however, have had a harder time of it. Of the sixty or so species on the mainland coast, only twenty-seven have successfully immigrated to Vancouver Island. The list of absentees includes Snowshoe Hares, Coast Moles, chipmunks, North American Porcupines, coyotes, Red Foxes and Grizzly Bears, to name a few.

Many species that did manage to colonize Vancouver Island and Haida Gwaii have evolved in isolation from their mainland populations long enough to be recognizably different. A few are now considered separate species, including the Vancouver Island Marmot. Most, however, have not crossed the species boundary and are classed as subspecies. Haida Gwaii is home to so many of these unique island forms that it has been called the Galápagos of Canada.

Perhaps the most interesting and complex example of island hopping is that of the deer mice. These big-eyed, big-eared mice are one of the commonest species of mammals throughout the province and are found on all but the tiniest coastal islands. Ian McTaggart-Cowan and other biologists exploring the small islands off Vancouver Island and Haida Gwaii noticed a strange pattern. There seemed to be two types of deer mice on these islands: a large, long-tailed type and a smaller, shorter-tailed type. The two types were never found together on the same small island. On some islands, such

as Triangle Island, the mice were almost three times the weight of mainland deer mice. Subsequent genetic studies have unravelled much of the mystery. There are two species, and both are found on the mainland as well as the islands. The big island mice are apparently Keen's Mice, a long-tailed species found high in the Coast Mountains and Cascade Range, and at lower elevations along the northern coast. Smaller island mice are the same species as the deer mice of lower elevations along the southern coast.

But why is there only one of either species on each small island, and how did it get there? The best guess is that the mice rafted from the mainland on large pieces of debris brought down the mountain slopes and into the sea by landslides. The first species to arrive on and populate an island could then exclude the other species by sheer weight of numbers.

Ancient Climate Change

We have seen how post-glacial immigration has shaped the geography of many plant and animal species in British Columbia. But climate also plays an important role in defining the distributions of species—some like it hot, some like it cold, some like it dripping wet and some can withstand months of drought. And climate is always changing. When most of us think of climate change, we think of the recent phenomenon of rapid, human-induced change, but over the long term the earth's climate has never been really static. And along with changes in climate come changes in the distribution of plants and animals. The range maps shown in bird and flower guides are not permanent and fixed but are always shifting.

Twelve thousand years ago, after the glaciers receded, the climate of the southern mainland was colder and drier than it is today; the coast received about 600 millimetres less rainfall annually than it does today. On the dry east side of Vancouver Island and the Gulf Islands, Helmeted Muskoxen, American Mastodons and Western

Bison roamed through an open landscape dotted with aspen groves.

Lodgepole Pine dominated the first forests to invade the southern mainland after the glaciers left. This species prevailed for about 2000 years, even as the climate became moister. But about 10,000 years ago, as the climate rapidly became warmer, Douglas-fir swept up the coast—even as far north as northern Vancouver Island—to replace

AMERICAN MASTODON

the pine. At this time, Sitka Spruce was also more common than it is today, but Western Hemlock played only a minor role; Western Redcedar was especially sparse.

In the southern Interior, cold steppe with sagebrush and grasses dominated, but scattered populations of Whitebark Pine, Western White pine, fir and spruce imply that the climate was cooler than today.

But the glaciers were shrinking from British Columbia because the Northern Hemisphere had entered a warm interval. Summer solar radiation reached a peak between 9000 and 10,000 years ago, when it was 8 per cent greater than today; winter solar radiation, in contrast, was 10 per cent less than it is today (see the box entitled "Planetary Cycles, Climatic Cycles" in Part I). With increased summer sunshine, the east Pacific subtropical high-pressure system expanded, intensifying summer drought. But the huge ice age glaciers continued to affect regional climate until they dwindled away, so the warmest period of the present interglacial period followed about 2000 years after the peak in solar radiation.

As a result, Interior summers 8000 years ago were much hotter and drier than they are today, and thus grasslands were much more

POLLEN: DUST OF THE AGES

How do we know what the climate was like thousands of years ago? Paleoecologists use a variety of clues to piece together the ancient landscape, but the most often used evidence comes from a very familiar source—pollen. Every spring and summer the pollen from a host of plants drifts on the warm breezes—giving some of us hay fever—and settles to the ground as a fine, dusty film. When pollen lands on a lake or bog, it sinks to the mud or peat on the bottom (have you noticed the golden rings of pine pollen around drying mud puddles?), where it is soon entombed by that year's sediments and debris. Preserved in this oxygen-poor environment, the pollen grains turn into millions of microscopic fossils, each of which can be identified. Summer after summer the pollen layers build up, continually recording the ecosystem around the pond or bog.

The translation of this record—which falls to a branch of science called palynology—is a difficult task and is often plagued by lack of information. Few lakes and bogs have been sampled and analyzed and, as Richard Hebda of the Royal British Columbia Museum writes, "Every lake tells a story—and adds another piece to the puzzle of the environmental history of British Columbia."

Pollen grains of alder, spruce and pine

extensive. East of the Coast Mountains, grass covered many south-facing mountain slopes, even in the far north. The lower timberline stood at around 1300 metres in the south, compared with about 500 metres today. The distribution of some Prairie and Great Basin plants and animals extended from Osoyoos all the way to central Yukon.

Conversely, wetlands and alpine areas were much less extensive than they are today. Many modern lake basins were dry or only held water ephemerally during the spring. Alpine tundra, so widespread in the mountains south of the ice during the glacial advances, dwindled and, on the lower, more isolated mountains, disappeared completely.

Between 7000 and 4000 years ago, the climate became moister and cooler. Along the southern coast, Western Hemlocks and then Western Redcedars expanded at the expense of Douglas-fir and Sitka Spruce. On southern Vancouver Island, Garry Oaks appeared and prospered—by 6000 years ago, they formed a continuous forest or woodland in the vicinity of Saanich Inlet. In the Interior, forests expanded and reclaimed most of the grasslands area. By 4500 to 3000 years ago grasslands had decreased to their minimum extent, and relatively modern conditions prevailed.

But even today you can see reminders of that warmer, drier time 8000 years ago—for example, relict grasslands on steep, south-facing slopes in northern British Columbia and a Douglas-fir stand on the slopes overlooking Stuart Lake, a lonely outpost of southern forest in the land of spruce.

The chronology outlined thus far is the story of southern British Columbia. The details of northern plant communities are less well known but can be inferred from two or three sites. In the far northwest, shrub tundra prevailed immediately following the retreat of the glaciers about 9250 years ago but quickly gave way to spruce woodland. There was a general increase in warmth and moisture until about 6100 years ago. Following a trend to a more semiarid climate about 4100 years ago, Subalpine Fir and then Lodgepole Pine joined the spruce forest about 3200 years ago.

In British Columbia, the last four millennia have been dominated by the coolest climate since the glaciers left. During this interval, cool summers and moist winters caused the expansion of mountain icefields and established the mature rain forests along the west coast. This interval culminated in the coolest period—sometimes called the Little Ice Age—which began in about 1650 and ended with a warming trend that started around 1850 and, with a minor decline in the middle of the last century, continues to this day.

Table 3 summarizes the climate changes of the last 16,000 years.

Mapping Today's Patterns

The patterns of living communities in British Columbia today are the product of interactions between geology, topography, climate, glaciation, colonization and the competition among species for space. Although these patterns may seem hopelessly complex—and, at finer levels, they may be—ecological zones that help make sense of British Columbia's biological diversity can be identified.

There have been several attempts to identify and map British Columbia's ecological zones, but only two systems are widely used today. Vladimir Krajina, a plant ecologist who worked at the University of British Columbia, developed a system of "biogeoclimatic zones," which are areas characterized by climatic factors and defined according to the tree species that dominate in climax forests (or grass species in climax grasslands, in nonforested areas) on average sites within them. These zones can be subdivided into subzones and subzone variants according to climate and vegetation.

Since biogeoclimatic zones help predict what plants will grow in certain areas and how much trees will grow, the provincial Ministry of Forests adopted and refined Krajina's system and mapped the province at the subzone variant level. Fourteen major zones are

TABLE 3: TIMELINE: CHANGING CLIMATE

Years ago	Events
16,000 to 13,000	Glaciers cover most of British Columbia. Tundralike ecosystems along parts of the east coast of Haida Gwaii.
13,000	Glaciers leave lowlands of southern mainland.
12,000	Climate cooler and drier than it is today. Forests in southern mainland dominated by Lodgepole Pine; the east side of Vancouver Island is open country with scattered Trembling Aspen groves.
10,000	Glaciers have left all the lowlands. Climate has become moister and now becomes rapidly warmer. Douglas-fir sweeps up the coast and replaces Lodgepole Pine; Sitka Spruce common, but Western Hemlock and Western Redcedar sparse. In the southern Interior, cold steppe with sagebrush and grass dominates; presence of subalpine trees implies that climate is cooler than at present.
10,000 to 9000	Summer solar radiation reaches peak—8 per cent greater than today, but huge glaciers still affect regional climate.
9250	Shrub tundra prevails in north, but spruce woodland dominates soon after.
8000	Warmest period of present interglacial period. Interior summers hotter and drier than at present. Grasslands much more extensive, but wetlands and alpine ecosystems much less extensive than today.
7000 to 4000	Climate becomes moister and cooler. Western Hemlocks and then Western Redcedars expand at the expense of Douglas-fir and Sitka Spruce.
6000	On southern Vancouver Island, Garry Oaks are more widespread than today.
4500 to 3000	Interior grasslands reach minimum extent; relatively modern conditions prevail there.
3200	Subalpine Fir and then Lodgepole Pine join spruce forest in north.
350 to 150	Climate cools, mountain glaciers expand during Little Ice Age.
150	Present warming trend begins.

	Alpine Tundra		Montane Spruce
	Spruce/Willow/Birch		Bunchgrass
	Boreal White and Black Spruce		Ponderosa Pine
	Sub-boreal Pine/Spruce		Interior Douglas-fir
	Sub-boreal Spruce		Coastal Douglas-fir
	Mountain Hemlock		Interior Cedar/Hemlock
	Engelmann Spruce/Subalpine Fir		Coastal Western Hemlock

YUKON

BRITISH COLUMBIA

NORTHWEST TERRITORIES

ALASKA

Fort Nelson

Fort St John

ALBERTA

Prince Rupert

Smithers

Prince George

Haida Gwaii

Fraser River

Kamloops

Vancouver Island

Vancouver

Nelson

Victoria

WASHINGTON

IDAHO   MONTANA

MAP 16

0   100 kilometres

now described (Map 16). An excellent summary of the characteristics
of all the biogeoclimatic zones is con-
tained in *Ecosystems of British Columbia*, published by the Ministry of
Forests.

MAP 16. BIOGEOCLIMATIC ZONES (*facing
page*). The biogeoclimatic zones of British
Columbia. *Courtesy of B.C. Ministry of Forests.*

Dennis Demarchi of the provincial Wildlife Branch took a different approach from Krajina's. He recognized a shortcoming of biogeoclimatic zones—in mountainous country they become complex,
sinuous strips following the contour lines along the fingerlike ridges.
But animal and plant populations do not necessarily experience the
landscape in narrow, zonal bands; they often use several zones. For
example, Western Redcedars in the southern Interior can live quite
happily in the Bunchgrass Zone, as long as they are beside a creek in
a cool, dark canyon, and they will also grow along the same creek in
the Interior Douglas-fir and Montane Spruce Zones upstream. In
one day a Grizzly Bear could traverse several biogeoclimatic zones
while moving from salmon stream to berry patch to ground squir-
rel colony. A band of California Bighorn Sheep may depend on a
snow-free slope of grass in the Bunchgrass Zone for winter forage,
a series of rock bluffs in the Ponderosa Pine Zone for escape terrain
and moist grasslands in the Interior Douglas-fir Zone for summer
pasture.

Demarchi wanted to map ecosystems the way an animal would
experience them. At the most general level, his "ecoregion" classification begins as Krajina's does—dividing the province into broad
climatic areas—but then it subdivides these areas according to
major physiographic units (such as plateaus, mountain ranges and
major valleys) as well as climate. There are five levels of organiza-
tion in the ecoregion classification: the first two—ecodomains and
ecodivisions—place British Columbia in a global context, whereas
the next three—ecoprovinces (Map 17, overleaf), ecoregions and
ecosections—relate portions of the province to the rest of North

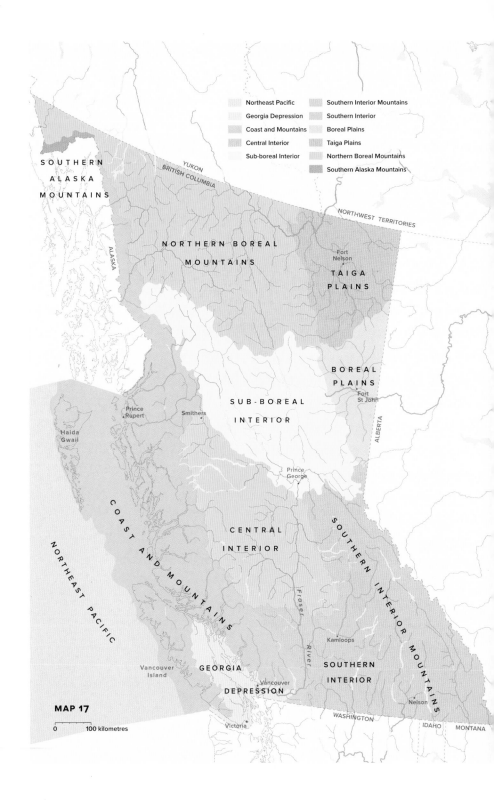

SOUTHERN
ALASKA
MOUNTAINS

Northeast Pacific
Georgia Depression
Coast and Mountains
Central Interior
Sub-boreal Interior

Southern Interior Mountains
Southern Interior
Boreal Plains
Taiga Plains
Northern Boreal Mountains
Southern Alaska Mountains

YUKON
BRITISH COLUMBIA

NORTHWEST TERRITORIES

NORTHERN BOREAL
MOUNTAINS

Fort
Nelson

TAIGA
PLAINS

ALASKA

BOREAL
PLAINS

Fort
St John

SUB-BOREAL
INTERIOR

Prince
Rupert

Smithers

ALBERTA

Haida
Gwaii

Prince
George

CENTRAL
INTERIOR

SOUTHERN INTERIOR MOUNTAINS

COAST AND MOUNTAINS

NORTHEAST
PACIFIC

Fraser
River

Kamloops

Vancouver
Island

GEORGIA

DEPRESSION

Vancouver

SOUTHERN
INTERIOR

Nelson

MAP 17

WASHINGTON

IDAHO
MONTANA

0   100 kilometres

Victoria

America or to other ecological areas within the province. Ecosection and ecoregion boundaries are primarily geographical barriers such as rivers

or lakes or the edges of different physiographic features, but the system also uses the boundaries of biogeoclimatic zones—for example, to separate highland areas from adjacent valleys.

British Columbia is far ahead of most other jurisdictions in North America in having its complex ecological regions described and mapped in such detail. And there is great advantage in having two different approaches to ecological classification—they can be tied together in a complementary way to produce a powerful, detailed method of describing the ecology of this province.

The Temporary Nature of Nature

Even though the regions of British Columbia can be described and mapped, the ecological map of the province is far from static. Although large-scale changes may not be readily apparent on human time scales, change is happening all the time—map boundaries are constantly shifting. Probably the most obvious changes humans can see in their lifetime are the great range expansions of some birds. Some of this avian immigration may be related to the climatic warming following the end of the Little Ice Age. As flying animals, birds can disperse widely and quickly and so offer the best examples of recent changes. For some birds, the extension of ranges northward over the last century has been remarkable.

The Wilson's Phalarope was first recorded in the province in 1922—just across the 49th parallel in the south Okanagan—but by the 1980s it had reached the marshes of the far northeast of British Columbia and the southern Yukon Territory. White-throated Swifts also put in their first historical appearance in British Columbia in the south Okanagan—in 1907. Now they career around rock bluffs

as far north as Williams Lake and the Clearwater River canyon. Most recently, Gray Flycatchers expanded northward from Oregon through the Ponderosa Pine forests of Washington State in the 1970s and reached the south Okanagan in the mid 1980s—where they are now a locally common species.

Conversely, some birds have spread west and south from the northeast. The Barred Owl was first recorded from the province along the Liard River in 1943, reaching the southern Interior (Wells Gray Park) by 1958 and the southern coast by 1966. We didn't see our first Barred Owls until the 1970s—but they are now found throughout the province and are one of the most dependable species to put on a loud hooting show on a spring night of owling.

facing page: The Barred Owl moved into British Columbia during the past century, reaching southern forests in the late 1950s and becoming commonplace there in the 1970s.

The question may be asked: if birds can fly anywhere they want, why don't all bird species move immediately into habitats they can make use of? One reason is that birds are quite conservative animals and resist moving into entirely new habitats. Gray Flycatchers, for instance, were formerly restricted to the sagebrush-juniper deserts of the Great Basin of the United States. In the late 1960s, a population of Gray Flycatchers took a liking to Ponderosa Pine forests along the east slope of the Oregon Cascades, perhaps because pine forests that have been selectively logged mimic the juniper woodland that the flycatchers are wired to look for. These flycatchers found the pine forests empty of fly-catching competition and quickly spread north.

Another question is whether the glaciers will return. Although the glaciers left the lowlands 12,000 years ago, there is no evidence that we are out of the Ice Age. The last two million years have been times of almost constant change, and if past patterns are any indication, we may only be living in a short interval of warmth amid a long period of ice. Looking at the past 200,000 years as a whole, many of the plant and animal species of southern British Columbia are more correctly seen as residents of California and Oregon visiting here temporarily during the warm season.

Time and Change

I (Sydney) often recall the June day I hiked into Berg Lake at the base of Mount Robson's north face. It was a cloudy day and the mountain was hidden by clouds that threatened snow in this high part of the Rockies. We tramped through forests clogged with last winter's snow on the steep face beside Emperor Falls. But upon emerging from the forests into the hanging valley south of the lake, I was awestruck—I suddenly felt I'd stepped back 12,000 years to the end of the Pleistocene glaciations. Tumbling Glacier hung suspended from the low clouds, and the river flats lay in front of us—a great expanse of gravel, Yellow Mountain-Avens and scattered conifers. It felt as though the glaciers had just left, and I honestly expected to see a small group of Woolly Mammoths amble away into the mists.

facing page: Tumbling Glacier flows out of the mists of Mount Robson's north face.

We don't often personally experience the profound changes that have taken place in our familiar landscapes over geological or even ecological time. With our short lifetimes and even shorter memories, we tend to view the world as static. Continents and mountain ranges stay put and islands remain offshore. We are not even aware of the movement of glaciers. Every once in a while a volcano erupts or an earthquake shakes us, but the landscape on a grand scale looks much the same. And on the time scale of years, living things seem static too. Polar Bears stay in the Arctic and sloths stay in the tropics. Sitka Spruce forests don't leave the coast and sagebrush steppes don't stray from the Interior. Seasons may come and go, but they seem to do so in a predictable fashion. Some Julys are hotter than others, but we can compare all of them with a reasonable average.

The ever-whirling wheel of change; the which all mortal things
doth sway...
EDMUND SPENSER, *The Faerie Queen*

But continents are moving and mountains are growing every
day—at 4 centimetres per year, the St. Elias Mountains have been
uplifted 4 metres in the last 100 years. The Rockies and the Columbia Mountains have lost a third of their glacial ice in the same time.
In the last fifty years, a third of the Kootenays' grasslands have
become forested. And all the evidence points to a rapid warming of
the world's climate, which undoubtedly is having a profound effect
on British Columbia's ecosystems. Change is the only real constant
in our world—British Columbia is being transformed in front of
our eyes.

---

facing page: A subvolcanic dyke, resistant
to the erosion that has claimed the land
around it, towers above a slope in the Ilgachuz Mountains, north of Anahim Lake.

APPENDIX

Geologically Special Places in British Columbia

(See map on page 142)

The Southern Interior

1. MOUNT ROBSON

Mount Robson, the highest of the Rocky Mountains, is a singular stand of glacially sculpted layered rock that reaches 3954 metres in elevation. For the best views, head along Yellowhead Highway (Highway 16) to 15 kilometres east of the junction with Highway 5 at Tête Jaune Cache. Mount Robson's south face rises 3000 metres from the valley bottom, exposing sedimentary strata that are of some of the older rock formations in the Rocky Mountains. These layers are pure limestones, quartz sandstones, shales and siltstones—all Cambrian in age. Later deposits—from the Ordovician, the Silurian, the Devonian, the Mississippian, the Permian and the Triassic—buried all of this. In Jurassic time, thrust-faulted and folded sheets of rock piled up in a growing heap: the offshore terranes scraped deep-water sedimentary rock off of the thin tapering edge of North America and pushed it up the ramp of the continental margin. Then in Cretaceous time, the slow-motion debris pile started to push at the shallow-water rocks, the thick limestones. They, too, were sheared off, folded, lifted and piled, and as the mass shortened east to west, it thickened enormously. This piling and thickening is what puts ocean sediments in the realm of the sky.

2. OKANAGAN VALLEY FAULT

As you approach this valley from either side along Highway 97, the view changes from unbroken hills and mountains to a gentle pastoral area, as though a piece of another country dropped into B.C.'s montane landscape. In a geological sense, this area indeed was dropped down—along a normal (top-down) fault, in which the rocks to the west slid off of the rocks to the east. The eastern rocks, formed deep in the earth's crust, once lay far below the piled-up terranes. They are banded gneisses, grey striped and crisscrossed by white granite and pegmatite (coarsely crystalline, quartz-feldspar-rich granite). Particularly good views of these are found near Vaseux Lake. The great Okanagan fault was active during Eocene time.

3. STEMWINDER MOUNTAIN AND HISTORIC HEDLEY (MASCOT) MINE

Along Highway 3 near Hedley, 38 kilometres east of Princeton— look for vistas of folded, striped limestone strata on the mountain face northeast of the highway. These pale cliffs are made up of Triassic limestone that was deposited at the edge of a volcanic basin— part of the Quesnellia terrane. The limestones are the "host" rock of the gold ores that were mined in the early twentieth century at Hedley. The village itself is worth a visit. There are historic hotels, a mining museum and tours to see the old mine buildings perched up on the ridge.

4. ZOPKIOS RIDGE GRANITE SCULPTURE

A few kilometres west of the Coquihalla Highway (Highway 5) summit you will find a prominent granite edifice north of the Zopkios rest area. This lovely and impressive mountain face is carved into Eocene granite of the Needle Peak pluton. Chunks of this rock lie near the road. Look at the large pink orthoclase (potassium feldspar) crystals in it if you get a chance. Otherwise, just admire the sloping slabs of the peak. As granites like this cool, they develop exfoliation joints—like the layers of an onion. The Coast Mountains here are largely made up of granite and its darker relatives, all formed by the melting of the down-going plate during Mesozoic to Eocene subduction of oceanic plates in the Pacific Ocean.

5. HOPE SLIDE

The scar from this huge landslide, seen from Highway 3 about 17 kilometres southeast of Hope, remains raw and open even decades after its fall on January 9, 1965. The slide, the largest recorded in Canada, killed four people and extinguished a small lake. Notice how the slide debris crossed the valley and slopped up on its southern side by over a hundred metres. The cause? Just the steepness of B.C.'s mountains, with some local complications due to fracturing and layering in the rocks that happened to be parallel to the slope. It is an enduring reminder that there is always the possibility of others like it.

6. HOPE CONGLOMERATE AND THE FRASER FAULT

The first road cuts on the northeast side of Highway 1 after crossing the Fraser-Hope Bridge leaving Hope show rock faces studded with large, round cobbles of granite, which distinguish them as conglomerate—rock made up of water-worn fragments of other rocks. There is a little parking area just past the road cut if you want a closer look. This conglomerate tells an important story about the geological history in this part of B.C. In Eocene time, a fault extended north–south through here, along what is now the valley of the Fraser River. North of Hope the river follows the trace of the fault, which explains why its course takes such a dogleg here as it breaks west towards Vancouver. Some ancient precursor of the

Fraser River also flowed through here. The size of the cobbles gives an idea of the strength of the current in it—much more energetic even than the modern Fraser, which at this point is depositing sand, silt and pebbles. If you look north up the Fraser Canyon from here you see ripples in the hills—points and dips. Each of the dips is an eroded-out strand of the ancient fault.

7. LYTTON DIKE SWARM

If you drive 5 kilometres east of Lytton along Highway 1, stop near the Skihist Provincial Park picnic site and look north across the Thompson River Canyon. You will see bluffs seamed with black stripes—igneous dikes—and this location has an unusually high concentration of them. In other areas you may see one or a few such dikes, but here this once-molten rock added over 40 per cent to the original volume of rock.

8. MR. MIKES MÉLANGE

At the southern end of the Cache Creek townsite, you will see a modest outcrop behind the buildings. The rock here is not in a nicely layered deposit—it is chaotic. Blocks of chert and argillite (mud-stone) are surrounded by a matrix of even muddier rock. Dating of microfossils has shown that the chunks and hunks are of many different ages, another indication of the mixed-up nature of this deposit. Geologists refer to this as a mélange, from the French word for "mix." Turbulent deposits like this form in deep-sea trenches and are also found in rock assemblages that formed along old subduction zones. Here, the mélange is part of the Cache Creek terrane, that remnant of exotic ocean floor that forms a strip in central B.C. The combination of mélange like this, and localities with exotic fauna such as *Yabeina* (see page 20), pegged the Cache Creek terrane as a tract of rock that must have travelled far to reach its present location. This outcrop became known as Mr. Mikes mélange, named for the restaurant that used to occupy the cinder block building in front of it.

9. BONAPARTE GOSSAN

About 12 kilometres north of Cache Creek, along Highway 97, stop at the pull-off on the west side of the highway, 4 kilometres north of the Pavilion turnoff, to view the Bonaparte gossan. This eye-catching rock display was created by hot fluids that came up a fault along the valley-side. The acidic fluids leached some elements out of the rock and added others—the reds and yellows are due to iron oxide, the pale overall colour to clays. Miners term these bright, iron-stained zones "gossan"—derived from the Cornish word for blood.

10. SERPENTINITES NORTH OF CACHE CREEK

In the few kilometres north of the Bonaparte gossan overlook, hill-slopes to the east show glistening dark-green fragments of serpentinite, a strange geological material. It is mostly one mineral, a hydrated magnesium-iron silicate. It originated not in the earth's crust but below it, where, in its unhydrated forms (olivine and pyroxene), it makes up most of the earth's mantle. Since the earth's crust is generally between 5 and 45 kilometres thick, mantle rocks at the surface are rare. The serpentinite here belongs to the Cache Creek terrane, a piece of ocean that was caught up in collisions between the ancient continent of North America and its fringing island arcs. The resultant forces scooped up the ocean floor and shoved it on top of the crust beside it. As the mantle at its base rose, water from surrounding rocks entered it and changed its crystalline minerals to the slick, greasy serpentinite that makes up these hills.

11. CHASM PROVINCIAL PARK

From Highway 97, turn off to the east 15 kilometres north of Clinton. Drive 3 kilometres along the access road until the view opens up beside you. The Chasm is cut by the eponymous creek into Miocene basalts (see the box entitled "Granite and Basalt" on page 39) that

cover much of the Cariboo-Chilcotin Plateau. Different flows are stacked up on both walls of the creek, some with red soil layers between them.

The Bonaparte gossan, a vibrant scree painting in iron-tinted ochres and yellows.

The Lower Mainland and Sea to Sky Corridor

12. NORTH SHORE MOUNTAINS

The twin peaks of the Lions are as iconic to Vancouverites as the Manhattan skyline is to New Yorkers. This is the southern edge of the Coast Plutonic Complex, a range dominated by large bodies of granites and their relatives (granodiorite, tonalite, diorite) that extends over 1000 kilometres, from here to southeastern Alaska. It arose during a long period of sustained subduction of the plates that made up the Pacific Ocean floor, under the western margin of the North American continent as it stood from Late Jurassic time. Just prior to that,

The canyon rim of the Chasm: basalt flows layered like a cake, with some bright red soil horizons in between.

most of the large terranes—Quesnel, Cache Creek, Stikine, Wrangellia and Alexander—had been added to the margin, building it out to its present location. For a closer look, head up the Sea to Sky Highway (99) towards Whistler.

13. STAWAMUS CHIEF

As you follow Highway 99 out of West Vancouver and descend towards Squamish, you come around a corner and suddenly see the impressive Stawamus Chief, a vast pillar of granodiorite beloved by rock climbers the world over. The granodiorite that makes it up is about 100 million years old (mid-Cretaceous). During this period (110 to 80 million years ago), most of the granitic material of the Coast Mountains was emplaced as rising magmas—molten mass after molten mass of it, shouldering aside the layered rocks of the crust, cooling as bulbous plutons. So in that sense Stawamus Chief

is typical. Its lacking of jointing—the term for cooling cracks and other late fractures that develop in granites—is what sets it apart as such a monolithic, unfractured edifice.

The face of Stawamus Chief, built to last from granodiorite with few fractures. Glaciers excavated the more friable rock around it and polished its dome.

14. MOUNT GARIBALDI AND THE BLACK TUSK

An impressive peak of rock and ice, Mount Garibaldi, visible from Highway 99 northeast of Squamish, first erupted between 300,000 and 200,000 years ago, but a series of major eruptions in the last 50,000 years built it into a classic stratovolcano. These eruptions occurred when the Squamish Valley was filled with a large river glacier that flowed out into what is now Howe Sound. The western flank of the mountain was built on top of this glacier, and as the ice melted about 13,000 years ago, the western side of Mount Garibaldi collapsed in a catastrophic landslide.

The Black Tusk in Garibaldi Park (seen here from the east) is the resistant core of a small volcano.

The same volcanic episode that built Mount Garibaldi also resulted in the Black Tusk, a ragged wedge of black basalt. This curious jet-black spire, visible on the eastern ridgeline from Highway 99, once formed the core of a small volcano, probably a cinder cone. The cinders weathered away, leaving this remnant.

Vancouver Island and Gulf Islands

15. GULF ISLANDS

From above, these islands in the southern Strait of Georgia look very different from the land around them. They are made up of parallel ridges and valleys with an overall curve convex to the southwest; in detail, they are wispy, fringed on their northwest and southeast sides by many points and bays. Most of the Gulf Islands are underlain by Late Cretaceous sedimentary rocks of the Nanaimo

Group—named for the Vancouver Island city. These strata are of thick, hard beds of conglomerate and sand-stone on one hand, which form cliffs,

Pillow basalts at the entrance to Witty's Lagoon near Tower Beach, now exposed on the shore of the same ocean in which they once formed an island.

ridges and points, and soft shales and thin sandstones on the other hand, which lie under the deep agricultural valleys and the harbours. The whole pile is tilted towards the northeast. It is also cut into pieces and piled together by thrust faults like the Fulford Fault, which runs through Fulford Harbour.

Nanaimo sandstones of the Gulf Island coasts feature a beautiful lacy texture, a result of holes of several sides growing together in the rock. Wave spatter leaves droplets on the rock, which dissolve the limy matrix of the sandstone. The resulting hole is more receptive to water, which in turn dissolves more of the stone, making an even larger chamber that can hold more water—a feedback loop over millennia.

16. BASALT PILLOWS IN WITTY'S LAGOON

Witty's Lagoon Park in Metchosin is well known among local geologists for its distinctive basalt formations. Basalt most often cools as flows, or spouts as cinders. But under the right conditions, under water, it can form a series of large, smooth globular shapes—termed "pillows" because of their appearance—and these can be seen at Tower Beach in Witty's Lagoon, ranging in size from 30 centimetres to several metres across. These basalts are Eocene in age, a little over 50 million years old. They formed as part of a chain of offshore islands that now makes up the Washington and Oregon Coast Ranges—the Crescent terrane. The island chain was then shoved up against the nearby continental margin, adding yet another piece of land.

17. SOOKE POTHOLES AND THE LEECH RIVER FAULT

Sooke Potholes Park is accessed via the Sooke River Road 5 kilometres north of Highway 14. The potholes are developed in basalt of the Crescent terrane. Five kilometres north of the park's campground, as you approach the vanished logging community of Leechtown, the topography changes dramatically from a canyon to an open river valley. The Leech River, which joins the Sooke River near here, comes in from the west-southwest following the trace of the Leech River Fault. The bedrock changes abruptly across the fault. Instead of boldly outcropping basalt, there are soft grey schists, seen mainly as cobbles along the riverbed. This is the Leech River schist, a panel of metamorphic rocks caught between the Crescent terrane and Wrangellia, the main bedrock of Vancouver Island.

18. BOTANICAL BEACH

This beach, located off Highway 14 near Port Renfrew, is renowned for its tide pools full of bright green anemones and deep purple sea urchins. To find it, turn left on Cerantes Road and drive 3 kilometres

to the parking area. The tide pools are developed in soft sandstone of the Sooke Formation, which, at

Water-carved basalt walls in the canyon of the Sooke River, viewed from the Galloping Goose trail, Sooke Potholes Park.

about 25 million years old, is very young as B.C. rock strata go. But the sandstone rests on top of a very different bedrock, composed of grey schists and sandstones standing up on their edges, cut by light-coloured granitic dikes. These strongly deformed rocks were deposited, buried, then deformed and metamorphosed, then uplifted and eroded down to the surface that lies underneath the Sooke sandstone. Later, that surface once again sank below sea level, and the sand was laid down on top of it. The base of the Sooke sandstone is what is known as an angular unconformity—*angular* because the layering in the Leech River strata has a different orientation than the Sooke beds, and *unconformity* because the two units do not conform to each other. The first recognition of angular unconformities in the 1780s—and understanding that different rock sequences had their own turbulent stories—proved to be a revelation in the study of the earth.

19. HORNE LAKE CAVES

Unbeknownst to most of its human residents, Vancouver Island boasts over 1000 caves, most of them in the Triassic Quatsino Limestone. The Horne Lake Caves are the most spectacular and easily accessible. To reach them drive 1 hour north of Nanaimo on the Island Highway (Highway 19), turn left at the Horne Lake Provincial Park sign (Exit 75) and drive 12 kilometres to the park. Horne Lake Caverns (www.hornelake.com) offers guided tours. Call 250-248-7829 for more information.

The Northern Interior

20. NISGA'A LAVA BEDS

Canada's last volcanic eruption occurred on Nisga'a land approximately 250 years ago, and you can see its aftermath 100 kilometres north of Terrace on the Nisga'a Highway. The lava destroyed everything in its path, sparked fires in the surrounding forests and covered two Nisga'a villages. More than 2,000 people perished. The vast lava beds still dominate the Nass Valley, filling it with rubble and pahoehoe (smooth) basalt with tipped blocks of glassy surface. Vegetation is just beginning to establish itself, in contrast to the dense rain forests around it.

21. STIKINE RIVER CANYON

A gravel road leads west for a hundred kilometres from Dease Lake on Highway 37 to Telegraph Creek along the lower Stikine River. Just to the south is Mount Edziza, northern B.C.'s most famous volcano, which erupted during the last ice age. Edziza flows blocked the Stikine River and forced it to cut a narrow, steep-sided new canyon.

The road clings to the sides of this canyon at some points. Columnar basalts are exposed on either side of the river. At the confluence with the Tuya River, a cliff shows radiating columns. These unusual cooling patterns form in lava tubes—linear caves that form as the last lava drains out of a cooling flow and that can later be filled with another flood of lava. Similar features form where lava forces its way in sinuous paths under glaciers.

Nisga'a lava beds near Gitwinksihlkw in the Nass Valley. The top of the basalt is made up of rubble: flow of the still-liquid lava ruptured the cooling crust and carried it along in a rolling, grinding jumble.

22. CASSIAR AND THE SYLVESTER ALLOCHTHON

About 2 hours' drive south of the Yukon border, Highway 37 passes the abandoned mining town of Cassiar in the heart of the Cassiar Mountains. The lower parts of the mountains are light-coloured limestone and dolomite. Their tops, though, are of black volcanic

The Sylvester allochthon near Cassiar. Dark ocean-bottom basalts and black sedimentary rocks deposited in deep, anoxic waters, thrust on top of platformal, shallow-water limestone and dolostone of the Cassiar platform.

and deep-water sedimentary rock. The upper unit, part of the Sylvester allochthon (*allo* meaning "strange" and *chthon* meaning "earth"), is the floor of a late Paleozoic ocean, with rocks about 360 to 270 million years old. It was thrust on top of the platformal limestones during Jurassic-Cretaceous compression as North America moved west to collide with its fringing continental fragments and island arcs. The contrasting shades of the underlying platformal and overlying oceanic rocks fit the dramatic story of their juxtaposition by mountain-building forces.

23. FOLDED MOUNTAIN
Limestone layers in the face of this mountain, visible from Highway 97 (Alaska Highway) 9 kilometres west of Toad River, form an

overturned V-shape. Folded Mountain is a beautiful example of how even thick rock strata can be folded under the right conditions. In order to fold, rocks need to be warm—and deep burial helps satisfy that condition. Also, force must be applied to them very slowly, over millions of years, so that they do not rupture and fault. Thinner strata fold more easily than thicker strata, limestone more easily than basalt. The Rocky Mountains, of which Folded Mountain is a part, are called a "fold and thrust belt"—part folded, part shuffled and stacked on thrust faults. Whether folding or faulting was more important in a given layer at a given time depended on local conditions. The overall effect, though, is of wave after wave of solid rock, as if a stone tsunami had washed up on the edge of the continent.

APPENDIX MAP. GEOLOGICALLY SPECIAL PLACES (*overleaf*).

APPENDIX MAP

0   100 kilometres

Organizations

NATURE CANADA
75 Albert Street, Suite 300
Ottawa, ON K1P 5E7

BC NATURE
1620 Mount Seymour Road
North Vancouver, BC V7G 2R9

**NORTHERN BRITISH COLUMBIA
PALEONTOLOGICAL SOCIETY**
c/o The Exploration Place
P.O. Box 1779, Prince George, BC V2L 4V7
www.theexplorationplace.com

**THOMPSON NICOLA
PALEONTOLOGICAL SOCIETY (TNPS)**
c/o Department of Geology
Thompson Rivers University
Kamloops, BC V2C 5N3
http://www.tru.ca/science/
geology/tnps.html

**VANCOUVER ISLAND
PALEONTOLOGICAL SOCIETY (VIPS)**
P.O. Box 3142, Courtenay, BC V9N 5N4
www.vips-fossils.com

**VANCOUVER PALEONTOLOGICAL
SOCIETY (VANPS)**
Centrepoint P.O. Box 19653,
Vancouver, BC V5T 4E7
www.vcn.bc.ca/vanps

**VANCOUVER ISLAND
PALEONTOLOGICAL MUSEUM
SOCIETY (VIPMS)**
151 West Sunningdale,
Qualicum Beach, BC V9K 1K7

**VICTORIA PALAEONTOLOGY
SOCIETY (VICPS)**
406–1155 Yates Street,
Victoria, BC V8V 3N1
www.vicpalaeo.org

**PEACE REGION
PALEONTOLOGY SOCIETY**
P.O. Box 1348,
Tumbler Ridge, BC V0C 2W0

For Further Reading

GENERAL

Farley, A.L. 1979. *Atlas of British Columbia: People, Environment and Resource Use.*
Vancouver: University of British Columbia Press.

Goward, T., and C. Hickson. 1995. *Nature Wells Gray.* 2nd ed.
Vancouver: Lone Pine.

Scudder, G.G.E., and N. Gessler, eds. 1989. *The Outer Shores.* Skidegate, BC:
Queen Charlotte Islands Museum Press.

Weston, J., and D. Stirling, eds. 1986. *The Naturalist's Guide to the Victoria
Region.* Victoria: Victoria Natural History Society.

GEOLOGY

Armstrong, J.E. 1990. *Vancouver Geology.* Edited by C. Roots
and C. Staargaard. Vancouver: Geological Association of Canada,
Cordilleran Section.

Briggs, D.E.G., D.H. Erwin and F.J. Collier. 1994. *The Fossils of the Burgess
Shale.* Washington, DC: Smithsonian Institution Press.

Clague, J., and R. Turner. 2003. *Vancouver, City on the Edge: Living
with a Dynamic Geological Landscape.* Vancouver: Tricouni Press.

Gabrielse, H., and C. Yorath, eds. 1991. *Geology of the Cordilleran
Orogen in Canada.* Ottawa: Geological Survey of Canada.

Gadd, B. 2009. *Canadian Rockies Geology Road Tours.* Canmore, AB:
Corax Press.

Gadd, B. 2009. *Handbook of the Canadian Rockies*. 3rd ed. Canmore, AB: Corax Press.

Gottesfeld, A. 1985. *Geology of the Northwest Mainland*. Kitimat, BC: Kitimat Centennial Museum.

Gould, S.J. 1989. *Wonderful Life: The Burgess Shale and the Nature of History*. New York: W.W. Norton & Co.

Harris, C. and H. Rhenisch, 2010. *Motherstone: British Columbia's Volcanic Plateau*. 108 Mile Ranch, BC: Country Light Publishing. http://www.chrisharris.com

Ludvigsen, R., ed. 1996. *Life in Stone: A Natural History of British Columbia's Fossils*. Vancouver: University of British Columbia. Press.

Ludvigsen, R., and G. Beard. 1994. *West Coast Fossils: A Guide to the Ancient Life of Vancouver Island*. Vancouver: Whitecap.

Mathews, W., and J. Monger. 2010. *Roadside Geology of Southern British Columbia*. Victoria: Heritage House.

Mathews, W.H. 1975. *Garibaldi Geology*. Vancouver: Geological Association of Canada, Cordilleran Section.

Nelson, J.L., and M. Colpron. 2007. Tectonics and metallogeny of the Canadian and Alaskan Cordillera, 1.8 Ga to present. In Goodfellow, W.D. (ed.), *Mineral Deposits of Canada: A Synthesis of Major Deposit Types, District Metallogeny, the Evolution of Geological Provinces, and Exploration Methods*. Mineral Deposit Division, Geological Association of Canada, Special Publication, 5, 755–791.

Price, R.A., and J.W.H. Monger. 2003. *A Transect of the Southern Canadian Cordillera from Calgary to Vancouver*. Geological Association of Canada, Cordilleran Section.

Roed, M.A. 1995. *Geology of the Kelowna Area and Origin of the Okanagan Valley, British Columbia*. Kelowna, BC: Kelowna Geology Committee, Okanagan University College.

Turner, R.J.W., R. Franklin, M. Ceh, C. Evenchick, N. Hastings, N. Massey and P. Wojdak. 2007. *Northern British Columbia Geological Landscapes Highway Map*. Geological Survey of Canada Popular Geoscience 94E; BC Geological Survey Geofile 2007-1. Available on the BC Geological Survey website and in some tourist information offices.

Yorath, C.J. 1997. *How Old Is That Mountain?* Victoria: Orca.

Yorath, C.J., and H. Nasmith. 1995. *The Geology of Southern Vancouver Island: A Field Guide*. Victoria: Orca.

Yorath, C.J. 1990. *Where Terranes Collide*. Victoria: Orca.

ECOLOGY

Meidinger, D., and J. Pojar, eds. 1991. *Ecosystems of British Columbia*. Victoria: B.C. Ministry of Forests.

Pielou, E.C. 1991. *After the Ice Age: The Return of Life to Glaciated North America*. Chicago: University of Chicago Press.

WEBSITES

BC Geological Survey
www.empr.gov.bc.ca/MINING/GEOSCIENCE/Pages/default.aspx

British Columbia Lapidary Society
www.lapidary.bc.ca/

Geological Survey of Canada: Educational sites
http://gsc.nrcan.gc.ca/education_e.php

Index

Bold page numbers indicate captions to maps and illustrations, and map legends.

Photo and Illustration Credits